SCIEN MATTERS

Nuclear Power

*prepared for the Course Team
by Malcolm Scott and
David Johnson*

Science: A Second Level Course

The S280 Course Team

Pam Berry (Text Processing)

Norman Cohen (Author)

Angela Colling (Author)

Michael Gillman (Author)

John Greenwood (Librarian)

Barbara Hodgson (Reader)

David Johnson (Author)

Carol Johnstone (Course Secretary)

Hilary MacQueen (Author)

Isla McTaggart (Course Manager)

Diane Mole (Designer)

Joanna Munnelly (Editor)

Pat Murphy (Author)

Ian Nuttall (Editor)

Pam Owen (Graphic Artist)

Malcolm Scott (Author)

Sandy Smith (Author)

Margaret Swithenby (Editor)

Jeff Thomas (Course Team Chair and Author)

Kiki Warr (Author)

Bill Young (BBC Producer)

External Assessor: John Durant

The Open University, Walton Hall, Milton Keynes, MK7 6AA.

First published 1993.

Edited, designed and typeset in the United Kingdom by the Open University.

Printed in the United Kingdom by Eyre & Spottiswoode Ltd, Margate, Kent.

ISBN 0749251026

This text forms part of an Open University Second Level Course. If you have not enrolled on the course and would like to buy this or other Open University material, please write to Open University Educational Enterprises Ltd, 12 Cofferidge Close, Stony Stratford, Milton Keynes, MK11 1BY, United Kingdom. If you would like a copy of *Studying with the Open University*, please write to the Central Enquiry Office, PO Box 200, The Open University, Walton Hall, Milton Keynes, MK7 6YZ, United Kingdom.

1.1

5195C/s280nucli1.1

Contents

I Introduction and aims

I.I Introduction

A *Times* editorial from 1842 stated:

> *We prefer to take our chance of cholera and the rest than to be bullied into health. England wants to be clean, but not cleaned by Chadwick.*

What do you think this editorial refers to? It refers to a proposal by Edwin Chadwick, in a report published in 1842, to build an underground sewerage system in London. Up to that time, London only had open drains, which went down the centre of the streets, and which eventually discharged into streams and rivers. This, in itself, sounds fairly horrifying to us now, but the problems that this gave rise to were compounded by the fact that there was no mains water supply. Instead, water was obtained from a myriad of wells, to which people went to draw their water. Not surprisingly, the water supplies frequently became contaminated with sewage, with the result that there were epidemics of water-borne diseases, like cholera and typhoid. Indeed, Queen Victoria's husband, Albert, died of typhoid, at the age of 42.

Against this background, why did *The Times* oppose what seems to be an extremely sensible proposal? To start with, any improvements in public health would have to be funded from taxes. Furthermore, the main burden of any additional taxes fell on the better off, who, in general, did not live in the crowded conditions that gave rise to the worst effects.

Additionally, in those days, the degree of state intervention in everyday life was very limited. For example, education was not compulsory, there were no state pensions and there was no health care. *The Times* was therefore reflecting both the general political norms of the day and at the same time protecting the financial interests of its readers.

This example is not exceptional: almost every technological and scientific innovation meets *some* opposition before it becomes accepted. The basis for such opposition may be straightforward; for example, the interests of a section of society are threatened, as in our example. Or it might be complex, involving a wide range of issues, such that advantages and disadvantages may need to be balanced. Nuclear power, of course, falls into this second category.

Even if an innovation is accepted at first, there is no guarantee that this acceptance will continue. For example, problems may arise which were not foreseen, or which were initially ignored, or were regarded as being of minor significance. Alternatively, the ideas, ideals and values that prevailed at the time the innovation was introduced may change, for a whole host of reasons. Such re-evaluations are usually not 'once and for all' judgements, but are continuous, with views changing as new facets of the issue are revealed. Again, this is the case with nuclear power.

To start thinking about nuclear power, consider your own views on it by undertaking Activity 1.1.

Activity I.I

Are you for, against, or ambivalent on the issue of whether nuclear power stations should be used at all? You may have little knowledge of nuclear power at this point,

---

but, nevertheless, write down your raw opinions about this question, along with a brief outline of the reasoning that lies behind them. One reason why we are asking you to do this is that when you have finished studying this book, you should find it interesting to see if either your opinions, or their basis, have been changed by the experience.

Even if *you* are ambivalent on the issue of nuclear power, the majority of people answering such a question would probably have quite definite views. However, as the answer to Activity 1.1 implies, those views may well have been dictated by a particular aspect of nuclear power, and be based on different levels of understanding. Yet in topics as complex as nuclear power, the need to make judgements based on a full appreciation of the issues involved would seem to be essential. Indeed, the following quote shows that the authors of the Sixth Report of the Royal Commission on Environmental Pollution (the 'Flowers Report') published in 1976 had no doubts about this.

> *521. We have explained our reasons for thinking that nuclear development raises long-term issues of unusual range and difficulty which are political and ethical, as well as technical, in character. We regard the future implications of a plutonium economy as so serious that we should not wish to become committed to this course unless it is clear that the issues have been fully appreciated and weighed; in view of their nature we believe this can be assured only in the light of wide public understanding.*

But what constitutes 'wide public understanding'? The basis for understanding is *information*, which must be readily available, so that those who are interested can use it to form *their* opinions. In this respect, nuclear power has had a particularly poor record. Thus, Warren Newman, Communication Director of AEA Technology [the (United Kingdom) Atomic Energy Authority], addressing the XIIth Public Relations World Congress in June 1991 said:

> *The [nuclear] industry overpromised what it could deliver; it was condescending when it told people not to worry and trust the experts; and excessive secrecy fuelled public anxiety.*

(Reported in *Atom*, July/August 1991)

Access to information, then, is a prerequisite for making an informed opinion. But what kind of information do we need? Let us list some of the issues involved in evaluating the role of nuclear power to see what is required. They are:

(i)   the cost of the electricity produced;

(ii) the risks arising from nuclear accidents;

(iii) the disposal of radioactive waste;

(iv) the relationship between nuclear power and nuclear weapons.

At first glance the first of these, the question of cost, would seem to be the most straightforward. But is it? One aspect of nuclear power which distinguishes it from other sources of energy is that both its waste products and the plant (nuclear reactor) at the end of its life are radioactive. This means that the wastes and decommissioned plant have to be safeguarded for hundreds of years, and this adds to the total cost of nuclear power. But the cost of these safeguards has not yet been established. For example, a report in *The Daily Telegraph* of 25 June 1991 regarding proposed changes in the method of safeguarding old reactors said:

> *Nuclear Electric believes the new programme will cut total decommissioning costs for the 13 stations it operates from earlier estimates of £3.9 billion to about £2.1 billion.*

Thus, even the apparently straightforward issue of cost depends on the assumptions that are made, so that there is immediately scope for differences of opinion. Once one starts to look at the other issues, the scope for such differences of opinion increases. Obtaining an informed opinion on nuclear power is therefore unlikely to be straightforward, as the following extract from the Flowers Report suggests.

> *499. It seems that nuclear power has in some ways become the whipping boy for technological development as a whole... The environmentalist tends to see those in the [nuclear] industry as being so committed to furtherance of their technology as to be wilfully blind to its dangers to the world. Those within the industry, many no doubt sustained by the thought that they are thereby making an essential contribution to the well-being of mankind, tend to see the environmentalists as people opposed to all technology who are prepared to denigrate their work on the basis of drummed-up and nebulous fears of future catastrophes... The arguments of both deserve to be heard with greater mutual understanding.*

However, although *some* of the recent opposition to nuclear power may be prompted by an opposition to technology itself (which need not, of course, invalidate such opposition) concern over the spread of nuclear power predates such considerations. This concern centred on the relationship between nuclear reactors and nuclear weapons. The initial development of nuclear reactors during, and immediately after, the Second World War was stimulated by the desire to produce plutonium for use in nuclear weapons; indeed, the atomic bomb dropped on Nagasaki in 1945 was made from plutonium. Since all reactors fuelled with uranium produce plutonium during their operation, this *potential* link with nuclear weapons is always present, and inevitably forms one of the considerations in any discussion of the role of nuclear power.

Paragraph 499 from the Flowers Report (quoted above) also points towards another problem, which is common to many environmental issues, namely that those opposing nuclear power and those who are responsible for producing it both believe in the value of what they are doing. Of course, those in the nuclear industry depend on its continuance for their livelihood, and this would suggest that they have a stronger motive for resisting its demise than the opposition have for promoting it. However, a desire to manipulate peoples' opinions can also be an extremely powerful motivator, even though the *material* benefit to the manipulators is less obvious, as countless religious wars testify.

Both sides in the nuclear debate will therefore seek to present any information relating to the issues in such a way as to support their point of view. Indeed, they would be abnormal if they did not! One must therefore be aware of the possibility of bias in the information from *both* sides. Dr Michael Flood, in the Friends of the Earth publication *The End of the Nuclear Dream* (1988), makes this point very clearly:

> *The public is bombarded from all sides with promotional material, some produced (largely at the public expense) by the Nuclear Electricity Information Group or by public relations departments within the nuclear industry; some produced by environmental groups and concerned individuals. The material rarely represents a balanced argument and is frequently misleading; statements are often unsupported, and sometimes unsupportable. This often serves to confuse rather than clarify issues.*

The UK nuclear power programme was started by the government in 1955, with the publication of the White Paper *A Programme of Nuclear Power*. Thereafter very substantial organizations grew up to run both the power production programme itself and the underlying research. As a consequence, when there are public inquiries into different aspects of the programme—for example, into whether or not to build a particular reactor (for example Sizewell-B or Hinkley Point-C) or processing plant (for example the THORP fuel reprocessing facility at Windscale)—the human and financial resources that the nuclear industry can draw on are very substantial. In contrast, the opposition groups often work with very small financial and human resources. Thus, the CEGB's costs for the Sizewell inquiry exceeded £10 million, whereas the budget for one of the main objectors, Friends of the Earth, was £120 000. Whether or not this imbalance created a bias in the outcome is a matter for debate; it certainly helped to ensure that justice did not *appear* to have been done.

1.2 What are the aims of this book?

We have identified the need to establish a body of informed opinion relating to any of the important issues confronting society. However, realization of this ideal is made difficult by a number of factors, including:

(i) the basic information may not be made available;

(ii) one issue may be linked with that of others;

(iii) bodies with a vested interest in the outcome may be biased in the information they give;

(iv) the resolution of many of the issues depends on consideration of ethical, economic and political factors.

Against this background, what are the aims of the Nuclear Power book of *Science Matters*? It must be said straightaway that it is *not* to come up with a 'considered judgement' on the issues we have identified. Our aim is to present the issues involved and the scientific background to them: it will then be up to *you* to make any judgement for yourself. Our aims can therefore be summarized as:

(i) to provide information on the scientific and technological background to nuclear power;

(ii) to indicate how this background relates to the issues involved;

(iii) to provide examples of the differences of opinion which exist on these issues.

The background science and technology involved is presented in Chapters 2 and 3. The next five chapters then consider five different aspects of nuclear power—biological aspects of radiation, the nuclear fuel cycle and waste disposal, accidents, economics and nuclear weapons proliferation. Finally, in Chapter 9, we take a look at the prospects for a form of nuclear energy which has not yet been commercially exploited, namely nuclear fusion. We also look more generally at the British energy scene, to provide a context for what you have learnt about nuclear power.

Consistent with the fact that we are not attempting to provide a considered judgement we shall not, for example, be making a detailed comparison of the costs of nuclear power with those of all the alternative ways of generating electricity. By focusing on the problems of reaching a considered opinion on nuclear power, we nevertheless hope to provide the right critical framework for you to undertake such comparisons yourself, or to appreciate those that have been made by others.

2 Nuclear power: the scientific background

In this chapter, we shall outline the key scientific concepts that govern the operation and safety of nuclear reactors. You will need to be patient because the relevance of some of the science may not become really clear until later chapters when we consider the issues involved in nuclear power in detail.

Box 2.1   Units of energy, power and interconversions

Energy comes in various forms, but in this book, we shall be especially concerned with energy transferred in the form of heat, and with **kinetic energy**, the energy a body possesses by virtue of its movement. The kinetic energy of a body is given by $\frac{1}{2}mv^2$, where m is the body's mass, and v is its velocity. The unit that energy is measured in is the **joule**, J. Thus, 1 J is the kinetic energy of a mass of 2 kg moving at 1 m s^{-1}.

▷   How is power related to energy?

▶   **Power** is the *rate* at which energy is transferred, for example when an electric fire transfers energy in the form of heat to its surroundings.

▷   What is the unit used to measure power?

▶   It is the watt (W), which corresponds to an energy transfer of 1 joule per second.

The watt is rather a small unit. Many domestic electrical appliances use powers measured in thousands of watts (**kilowatts**, kW), and electricity power stations normally have their outputs rated in millions of watts, that is, in **megawatts (MW)**.

▷   How does one measure the total electrical energy produced or used?

▶   In joules is the most sensible answer, but because the *total* energy is equal to the product of the power and the time for which it is being produced, the energy industry often uses the **kilowatt hour (kW h)**—that is, one kilowatt being produced or used for one hour, or 1 000 watts being produced for $60 \times 60 = 3\,600$ seconds. One kilowatt hour is therefore 3 600 000 J (3.6×10^6 J). Larger units (MW h, that is 10^3 kW h, and GW h, 10^6 kW h) may be used to measure the total energy produced by power stations.

▷   What is the unit used in selling electricity to domestic consumers?

▶   Pence per kilowatt hour (p kW h^{-1}). ∎

2.1 Atoms and nuclei

Many of the organizations responsible for overseeing the development of nuclear power have the words 'atomic energy' in their titles. In the UK, for example, we have the United Kingdom Atomic Energy Authority (UKAEA). But, as you know, all matter is composed of atoms. We could therefore equally well call the energy obtained by burning coal 'atomic energy'! So what is the difference between the atomic energy these organizations are interested in and that obtained from coal? To answer this question, you first need to be sure of your ideas about atoms and nuclei.

2.1.1 The structure of atoms and nuclei

At the heart of every atom is a compact core, the **nucleus**. All nuclei contain one or more **protons**, each of which carries one positive charge. It is the number of protons which determines the identity of the **element**. For example, hydrogen has one proton in its nucleus, sodium has 11, lead has 82 and uranium has 92.

The other particle which is in all but one nucleus is the **neutron**. The neutrons and protons are tightly bound together by a special **nuclear force**. The masses of protons and neutrons are very similar, and both are very small ($1.672\,62 \times 10^{-27}$ kg for the proton and $1.674\,93 \times 10^{-27}$ kg for the neutron), but the neutron is neutral; that is, it has no electrical charge. The protons and neutrons in a nucleus are known collectively as **nucleons**: a nucleon may be either a proton or a neutron.

To complete our picture of the atom, we have to add a third particle, the **electron**. This has a much smaller mass ($9.109\,53 \times 10^{-31}$ kg) than either the proton or the neutron—that is, 1 836 times smaller than the proton—and it carries a *negative* charge equal in magnitude to that of the proton. But the whole atom is electrically neutral, which means that the number of electrons in the atom must be equal to the number of protons in the nucleus.

The electrons can be considered to move around the nucleus at distances which are very large compared to the radius of the nucleus. Consequently, the nucleus only occupies a tiny fraction of the total volume of the atom. For example, the diameter of a sodium nucleus is approximately 6.8×10^{-15} m, whereas the sodium atom, that is, including the electrons, has a diameter of about 3.8×10^{-10} m. Using these figures, the volume of the sodium nucleus is about 10^{14} times smaller than that of the sodium atom!

Although the number of protons in the nucleus of a particular element is fixed, the number of neutrons may vary; each different combination of protons and neutrons is called an **isotope**. If we want to identify an isotope fully, we have to specify the number of protons *and* neutrons. We shall use two similar notations for identifying isotopes. Both require the chemical symbols for the element, and you will find these symbols in Appendix 1. The elements are listed there in alphabetical order; for each element the symbol, atomic number and relative atomic mass are given. In our first notation, isotopes are identified by using the element's chemical symbol, preceded by the number of protons (the **atomic number**, Z) as a subscript, and the total number of protons and neutrons (the **mass number**, A) as a superscript.

As an example of this notation, consider the case of uranium (with an atomic number of 92), which has two naturally occurring isotopes, one with 143 neutrons in each nucleus and the other with 146. The mass numbers of these two isotopes are thus 235 ($= 92 + 143$) and 238 ($= 92 + 146$). The symbol for uranium is U, and we would therefore write these isotopes as $^{235}_{92}\text{U}$ and $^{238}_{92}\text{U}$.

The second notation is simpler in that it leaves out the atomic number, which was included in the first notation: it simply gives the symbol for the element together with the mass number. For example, the two uranium isotopes just mentioned would be written as ^{235}U and ^{238}U, or as uranium-235 and uranium-238.

Question 2.1 This question tests your ability to use the two notations and Appendix 1 correctly. If isotopes of hydrogen, sodium and lead have 0, 11 and 126 neutrons, respectively, use the information given above to write these isotopes in the two shorthand notations.

2.1.2 Energy from nuclear reactions

Most of the *chemical* reactions that go on around us, like the burning of coal or natural gas in power stations and elsewhere, release heat to their surroundings. Using chemical notation, we can write the reaction between carbon, the main constituent of coal, and the oxygen in the air as

$$C(s) + O_2(g) \longrightarrow CO_2(g) \qquad (2.1)$$

The heat produced in a chemical reaction is usually expressed in joules (J). However, the energy from nuclear reactions is normally given in electronvolts (eV), and is quoted per reacting nucleus. So that we can compare chemical and nuclear energy, let us work out the chemical energy released in burning carbon in terms of eV per atom burnt.

Measurements show that when, in Equation 2.1, one atom of carbon reacts with one molecule of oxygen, the energy released is 6.54×10^{-19} J.

We now have to convert this into eV.

$$1\,eV = 1.602 \times 10^{-19}\,J, \text{ so that } 1\,J = \frac{1}{1.602 \times 10^{-19}}\ eV$$

Hence the energy released per atom of carbon burnt is

$$6.54 \times 10^{-19} \times 1/(1.602 \times 10^{-19})\,eV = 4.08\,eV$$

This figure is typical of the energy changes in chemical reactions: the energy released is no more than a few electronvolts per atom.

Let us now look at nuclear reactions. Consider a reaction in which a proton (the nucleus of the simplest isotope of hydrogen) combines with a neutron. Because the neutron has a mass number of one, and contains no protons, it is written $^1_0 n$. Thus, we have

$$^1_1 H + {}^1_0 n \longrightarrow {}^2_1 H + energy \qquad (2.2)$$

Equations for nuclear reactions such as Equation 2.2 must be balanced. This means that the sum of the superscript mass numbers of the particles must be the same on both sides of the equation; in this case that number is two. It also means that the sum of the charges on each particle (thus, zero for neutrons and one for each proton) must be the same. In this case that number is one, the atomic number (subscript) of hydrogen.

Nuclear reactions differ from chemical ones in that the elements and their isotopes can *change* as a result of the reaction. Here, for example, a new isotope of hydrogen, $^2_1 H$, is produced. It is sufficiently important to have its own name, **deuterium**, and is sometimes given the symbol D ($^2_1 D$).

Now let us turn to the energy released in this reaction. Einstein showed that mass, m, and energy, E, were related via the equation

$$E = mc^2 \qquad (2.3)$$

where c is the speed of light ($2.998 \times 10^8\,m\,s^{-1}$).

▷ If the mass is expressed in kg, and the speed is in $m\,s^{-1}$, then what unit will the energy be in?

▶ The energy unit will be $kg\,(m\,s^{-1})^2$ or $kg\,m^2\,s^{-2}$. However, $1\,kg\,m^2\,s^{-2} \equiv 1\,J$, so the unit is simply joules (J).

One significance of Equation 2.3 is that if mass is lost in a reaction, then its equivalent in energy will be released; conversely, if mass is created, then an equivalent amount of energy will have been used to create it. Einstein's relationship says, therefore, that mass and energy are *equivalent*; energy can be expressed in terms of its equivalent mass, and vice versa.

Let us now look at the **mass balance** for Equation 2.2. The mass of the neutron and proton were given earlier as $1.674\,93 \times 10^{-27}$ and $1.672\,62 \times 10^{-27}$ kg, respectively. The mass of the deuterium nucleus is $3.343\,58 \times 10^{-27}$ kg.

▷   What is the change in mass when the reaction in Equation 2.2 occurs?

▶   The sum of the masses of the neutron and proton is $3.347\,55 \times 10^{-27}$ kg, so that $(3.347\,55 - 3.343\,58) \times 10^{-27}$ kg, that is, $0.003\,97 \times 10^{-27}$ kg, is lost when this reaction occurs.

But we have said that mass that is lost is transformed into energy.

▷   What is the energy released, in MeV, when the reaction in Equation 2.2 occurs?

▶   The energy E released will be given by

$$E = mc^2$$

$$= (0.003\,97 \times 10^{-27}\,\text{kg}) \times (2.998 \times 10^8\,\text{m s}^{-1})^2$$

$$= 3.57 \times 10^{-13}\,\text{J}$$

(handwritten annotations: "mass lost" pointing to m; "c^2"; circle around J)

We now convert this into electronvolts:

$$E = \frac{3.57 \times 10^{-13}\,\text{J}}{1.602 \times 10^{-19}\,\text{J}}\ \text{eV}$$

$$= 2.23 \times 10^6\,\text{eV}$$

$$= 2.23\,\text{MeV}$$

So, in this nuclear reaction the energy released is of the order of one million times greater than for the chemical reaction in Equation 2.1. This highlights the second difference between chemical and nuclear reactions: the energy changes associated with nuclear reactions are far larger than for chemical ones.

Nuclear binding energy

In the example that we have been discussing, a deuterium nucleus was formed by combining a proton and a neutron, and we saw that 2.23 MeV of energy was released. If we were to build up other nuclei in the same way, by forming them from *separate* protons and neutrons, we would find that energy was released in each case; that is, *the mass of any nucleus is less than the sum of the individual masses of the protons and neutrons that it contains.*

However, we know that if mass is lost, energy is released. The total energy that is *released* when a nucleus is *formed* from its constituent neutrons and protons is called the **binding energy** of the nucleus. Alternatively, the binding energy may be regarded as the energy that must be *supplied* when a nucleus is *broken up* into its constituent protons and neutrons. Figure 2.1 clarifies this, again using the example of deuterium, for which, as we have seen, the binding energy is 2.23 MeV.

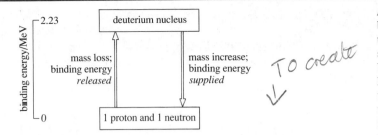

Figure 2.1 The concept of binding energy illustrated using the example of deuterium. The separate proton and neutron at the bottom are not bound together, and have no binding energy. When they combine to form deuterium, the mass of the deuterium nucleus is less than that of the neutron and proton from which it was formed. The mass loss is converted into energy, the so-called binding energy of the deuterium nucleus, which is *released*. This binding energy is 2.23 MeV (see text). Conversely, when the deuterium nucleus is broken up into its constituent neutron and proton, there is a mass increase. This disintegration of the nucleus can only occur if the binding energy of 2.23 MeV is *supplied* from the surroundings.

Binding energy is released every time a nucleon is added to a nucleus. The *average* binding energy released for each nucleon in the nucleus is equal to the total nuclear binding energy divided by the total number of nucleons.

▷ What is the average binding energy per nucleon for deuterium?

▸ In a deuterium nucleus there are two nucleons (one proton and one neutron). The average binding energy per nucleon for deuterium is (2.23 MeV/2), that is 1.115 MeV.

Question 2.2 The mass of the nucleus of the commonest isotope of oxygen, $^{16}_{8}O$, is 26.55297×10^{-27} kg. Using the information given earlier, calculate its binding energy, and the average binding energy per nucleon, in MeV. To do this you need to calculate the total mass of the individual protons and neutrons in oxygen-16, the difference between this and the mass of the oxygen-16 nucleus, and the energy-equivalent of this mass change.

You will have found from this question that the average binding energy per nucleon for oxygen-16 (mass number $A = 16$) is far greater than for deuterium (mass number $A = 2$), showing that the binding energy per nucleon is not a constant quantity. The way in which it differs from nucleus to nucleus is shown in Figure 2.2. We see that it increases with increasing mass number (although not smoothly) up to a value of A of about 60, where the binding energy per nucleon is about 9 MeV; it then decreases slowly.

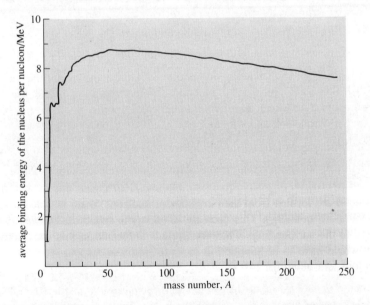

Figure 2.2 The variation in average binding energy per nucleon with mass number, A.

We shall see later that we can use the data in Figure 2.2 to explain why energy is released when nuclei of low mass number fuse together (*fusion* energy, discussed in Chapter 9), or when nuclei of high mass number split up (*fission* energy, discussed later in this chapter).

2.1.3 Nuclear stability

Can *any* combination of numbers of neutrons and protons exist as a nucleus? If we look at the numbers of each particle in the **stable isotopes**, (that is, those that do not break down by **radioactive decay**), then we find that there are, in fact, only a limited number of combinations possible. This can be seen from Figure 2.3, which plots the number of protons (Z) versus the number of neutrons ($A - Z$) for all the *stable* isotopes. The coloured line shows where the points would lie for equal numbers of protons and neutrons. You can see from Figure 2.3 that, for the low-mass elements (that is, with Z less than 6 or so), the numbers of protons and neutrons are approximately equal. However, as the nuclear mass increases, the proportion of neutrons to protons increases.

▷ What is the ratio of neutrons to protons in $^{23}_{11}$Na and in $^{208}_{82}$Pb?

▶ There are $(23 - 11)$ neutrons in the sodium-23 isotope, so the ratio of neutrons to protons in this isotope is $(12/11)$, that is, 1.09. There are $(208 - 82)$ neutrons in the lead-208 isotope, so the ratio is $(126/82)$, that is, 1.54.

Notice that the isotope of higher atomic number, $^{208}_{82}$Pb, has the higher ratio of neutrons to protons. This trend agrees with Figure 2.3 and has important implications for the fission process. Notice, however, that what Figure 2.3 shows most clearly is that the stable isotopes only occur in a narrow band. The unstable nuclei that fall outside this band either have too many neutrons (the ratio of neutrons to protons is too large) or too many protons (the ratio of neutrons to protons is too small).

Nuclei that are unstable undergo radioactive decay, changing their composition in order to become more stable. We shall be discussing this in the next section.

Summary of Section 2.1

1 Almost all of the mass of an atom resides in a highly compact nucleus consisting of protons and neutrons (*nucleons*). The neutron is neutral, but the proton carries a single positive charge.

2 The number of protons in a nucleus is the atomic number (Z). Each chemical element has a characteristic atomic number, which distinguishes its nuclei from those of all other chemical elements.

3 The proton and the neutron have similar masses. The mass number of a nucleus is the total number of nucleons it contains.

4 The different combinations of mass numbers and atomic numbers which are found in nuclei of the same element are called *isotopes*. Nuclear reactions involve the transformation of isotopes.

5 When nuclei are formed from their constituent nucleons, there is a decrease in mass. This mass decrease is converted into an amount of energy known as the *binding energy* which is released to the surroundings when the nucleus is formed in this way; conversely, this energy would have to be supplied in order to break the nucleus up into its constituent protons and neutrons.

6 The binding energy per nucleon increases with A up to about $A = 60$, and thereafter decreases slowly.

7 Nuclei that are stable (that is, do not undergo radioactive decay) have a neutron-to-proton ratio close to one at very low values of Z, but this ratio increases steadily to about 1.54 in the $Z = 80$–90 region.

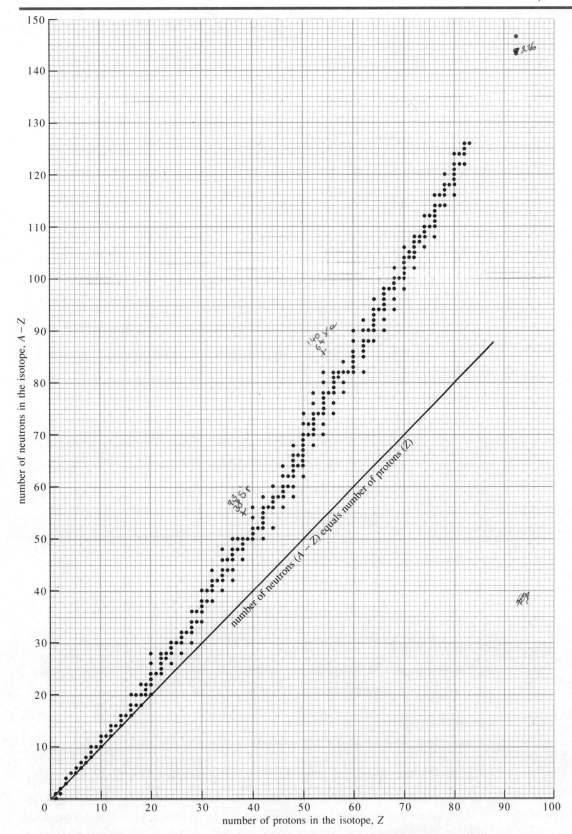

Figure 2.3   The variation in the number of protons and neutrons in the nuclei of stable isotopes. The positions of $^{235}_{92}U$ and $^{238}_{92}U$ are shown in colour; neither are completely stable, but both have very long half-lives. The straight line shows the positions that isotopes with equal numbers of protons and neutrons would occupy.

Question 2.3 To help you to check that you have understood the key points of Section 2.1, say whether the following statements are true or false.

T (a) Atoms consist of a nucleus, which contains protons and (with one exception) neutrons, surrounded by electrons.

F (b) The volume of atoms is not much larger than that of their nucleus.

T (c) Mass and energy can be considered to be equivalent; if mass is lost, energy is created, and if mass is created, energy is lost.

F (d) As the number of protons in a nucleus increases, the ratio of neutrons to protons needed to keep the nucleus stable decreases.

F (e) The nuclear binding energy is the energy released when a nucleus is broken up into separate, individual, neutrons and protons.

T (f) Energy is released every time a nucleon is added to a nucleus.

2.2 Nuclear decay — radioactivity

When unstable nuclei decay, they emit radiation or particles that can cause biological damage, and hence affect health. In this section we shall be considering the important characteristics of the different radioactive decay processes. In later chapters we shall use this information to study how radiation can affect health, and to see what the risks are from the use of nuclear power.

In order to guide yourself through this section, you should complete the following table as you go along.

Radioactive decay process	Particle(s) emitted	Nuclear change in	
		A	Z

2.2.1 The radioactive decay process

If nuclei do not have combinations of neutrons and protons which fall within the range of stable configurations shown in Figure 2.3*, what happens? The answer is that the nuclei are unstable, and undergo **radioactive decay**. Isotopes that undergo radioactive decay are called **radioisotopes**. The *way* in which radioisotopes decay depends on whether the nucleus has too many protons or too many neutrons for stability.

* Note that there are also some *unstable* isotopes within the range of stable nuclei shown in Figure 2.3.

β-decay

Let us first consider the situation in which there are too many neutrons. In this case, one of the neutrons can decay into a proton and an electron (a particle called an *antineutrino* is also produced, but as it has no charge and very little mass, it can normally be neglected).

If, for simplicity, we just write the neutron as n, the proton as p^+, and the electron as e^-, the decay of the neutron becomes:

$$n \longrightarrow p^+ + e^- \tag{2.4}$$

Notice that charge is conserved in the process: the charges of the proton and electron on the right are equal and opposite, so the overall charge is zero on both sides of the equation.

When the decaying neutron is part of a nucleus, the proton that is produced stays in the nucleus, but the electron is ejected from the nucleus with considerable kinetic energy. The process is then called **beta (β)-decay**, the ejected electron is called a **beta (β)-particle**, and a nucleus decaying in this way is called a **beta (β)-emitter**.

Because a neutron is replaced by a proton, a nucleus that undergoes β-decay will be transformed into a different nucleus. Let us consider an example.

▷ $^{14}_{6}C$ is unstable, and undergoes β-decay. What isotope will be formed when a $^{14}_{6}C$ nucleus decays in this way?

▶ The nucleus of carbon-14 has 6 protons, that is, $Z = 6$, and has $(14 - 6) = 8$, neutrons. One of these neutrons decays into a proton, so that the new nucleus will have seven neutrons and seven protons. From Appendix 1 you will see that the element with 7 protons in its nucleus (that is, with atomic number 7) is nitrogen; the new isotope will therefore be $^{14}_{7}N$ (nitrogen-14).

Thus, in β-decay the atomic number, Z, of the new isotope is one greater than that of the decaying one, but the mass number, A, is unaltered.

β-decay is one of the most important radioactive decay processes that radioisotopes produced in a nuclear reactor undergo; the reason why will be explained in Section 2.3.1.

Positron emission $\left(\beta^+\right)$

Now consider the case where the nucleus has too few neutrons (too many protons) to be stable. In this case, a proton can be transformed into a neutron, and a positively charged electron, e^+ (a **positron**):

$$p^+ \longrightarrow n + e^+ \tag{2.5}$$

The neutron stays in the nucleus, but the positron is emitted from it, often with considerable amounts of kinetic energy. Again, charge is conserved; there is one unit of positive charge on each side of the equation.

▷ What isotope is formed when sodium-22 ($^{22}_{11}Na$) undergoes radioactive decay by **positron emission**?

▶ One of the protons will be lost, so that the atomic number of the element formed will be $(11 - 1)$, that is, 10. However, the mass number, A, will not be changed, because a neutron replaces a proton in the new element. Looking at Appendix 1 you can see that the element with $Z = 10$ is neon. Neon-22 ($^{22}_{10}Ne$) is therefore the new isotope formed when sodium-22 undergoes radioactive decay by positron emission.

So, in positron emission, the mass number, A, is unaltered, but the atomic number, Z, of the new element is one less than that of the decaying nucleus.

Although radioisotopes decaying by positron emission have important applications in medicine, they are rarely produced in nuclear reactors. You will see why in Question 2.7.

α-decay

If a nucleus is unstable and has a high mass number ($A > 160$) and a high atomic number ($Z > 60$), it is possible for two neutrons and two protons to be emitted *together*. This combination is called an **alpha (α)-particle**, the process is called **alpha (α)-decay**, and the nucleus is often referred to as an **alpha (α)-emitter**.

The new nucleus therefore has two fewer protons and two fewer neutrons than the original one. Thus, in α-decay, the atomic number, Z, is reduced by two, and the mass number, A, is reduced by four. If you look at Appendix 1, you will see that the α-particle emitted, which has mass number 4 and atomic number 2, is a helium nucleus ($^{4}_{2}\text{He}$).

▷   What are the atomic number, Z, and mass number, A of the mercury (Hg) isotope that is formed when lead-184 ($^{184}_{82}\text{Pb}$) undergoes α-decay?

▶   The atomic number of the mercury will be two less than that of the lead isotope, so it will be 80. The mass number will be reduced by four, and will be 180. The new isotope will therefore be $^{180}_{80}\text{Hg}$.

We can write this example of α-decay as:

$$^{184}_{82}\text{Pb} \longrightarrow {}^{180}_{80}\text{Hg} + {}^{4}_{2}\text{He} \tag{2.6}$$

You will remember from Section 2.1.2 that the elements and isotopes in nuclear reactions can change, in contrast to the situation with chemical reactions. Equation 2.6 shows this very clearly, since one element undergoes radioactive decay to form two different elements.

Table 2.1 summarizes these three decay processes and, at the same time, allows you to check the table that you have completed on p.16.

Table 2.1   Summary of the characteristics of the principal radioactive decay processes.

Radioactive decay process	Particles emitted	Nuclear change in	
		A	Z
β-decay	β-particles (electrons)	0	+1
positron emission	positrons (positively charged electrons)	0	−1
α-decay	α-particles (helium-4 nuclei)	−4	−2

The three radioactive decay processes we have discussed so far are the most common ones. However, in the very high mass nuclei (such as uranium) it is possible for the nucleus to decay by splitting into two smaller nuclei with an associated emission of neutrons. When this happens to an isolated nucleus, it is called **spontaneous fission**; californium-252 is an example of an isotope that decays in this way. **Fission** is the general term applied to processes in which nuclei split into two parts. From our point of view the most important type is neutron-induced fission, which we shall be discussing in much more detail in Section 2.3.1.

Question 2.4 Below are listed three pairs of isotopes. In each case, the first isotope given is radioactive, and decays to form the second one. Say in each case what radio-active decay process is taking place.

(i) $^{15}_{8}O$ $^{15}_{7}N$ *positron emission.*

(ii) $^{235}_{92}U$ $^{231}_{90}$~~Ra~~ Th   *α- decay*

(iii) $^{60}_{27}Co$ $^{60}_{28}Ni$ *β - decay*

General comments on radioactive decay

When radioactive decay takes place, with the emission of α-particles, β-particles or positrons, new nuclei are formed and energy is released. The *way* in which this energy is released depends on the particular radioisotope concerned. In some cases, all the energy is given to the particles emitted, as kinetic energy. In the majority of cases, however, energy is also given off in the form of one or more photons of **γ-radiation**, also known as **gamma (γ)-rays**; these are a form of electromagnetic radiation emitted from a nucleus.

The α- and β-particles and γ-rays produced during radioactive decay all have the property that they can penetrate matter, though they vary enormously in their ability to do so. α-particles can be stopped by a piece of paper, or the human skin, whereas β-particles require a few millimetres of metal to absorb them. γ-rays, on the other hand, are very penetrating, particularly if they have a high energy (an MeV or so); as we shall see (Section 3.1.4), nuclear reactor shields comprise several metres of concrete for this reason.

Although we have yet to discuss their origins and properties in detail, you should note that neutrons, particularly high-energy ones (that is, having energies measured in MeV) are also very penetrating, and are like γ-rays in this respect. These differences are illustrated in Figure 2.4.

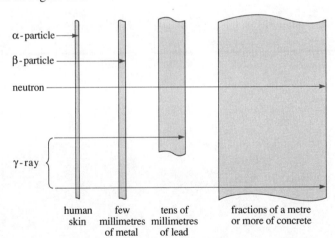

human skin   few millimetres of metal   tens of millimetres of lead   fractions of a metre or more of concrete

Figure 2.4 The penetrating power of different forms of radiation. γ-rays are usually stopped with lead shields from a few millimetres to tens of millimetres in thickness according to the energy of the radiation, but lead is expensive. Concrete is often used because it is cheaper, but a greater thickness is needed.

The absorption (stopping) of the particles or radiation emitted during radioactive decay results in heat being produced. The fact that heat is produced as a result of radioactive decay plays an extremely important role in nuclear reactor safety, as we shall see in Chapter 6.

Question 2.5 Plutonium-239 is a radioisotope produced in nuclear reactors, which can be used as a nuclear fuel, or in the manufacture of nuclear weapons. It is normally handled using rubber gloves while being viewed through a thin Perspex screen. What type of radioactive decay does this suggest is taking place? Use Appendix 1 to help you write the equation for this decay.

α particle emitted

$^{239}_{94}Pu \rightarrow {}^{235}_{92}U + {}^{4}_{2}He$

Radioactive decay chains

It is possible that, when a radioisotope decays, the new nucleus formed is *itself* unstable, and a further radioactive decay, or even a sequence of them, occurs. The initial radioisotope is then said to give rise to a **radioactive decay chain**.

A complex example arising from a naturally occurring radioisotope (uranium-238) is shown in Figure 2.5. This particular chain is important because one of the **daughter products** (as members of the decay chain are called) is the gas radon, $^{222}_{86}\text{Rn}$.

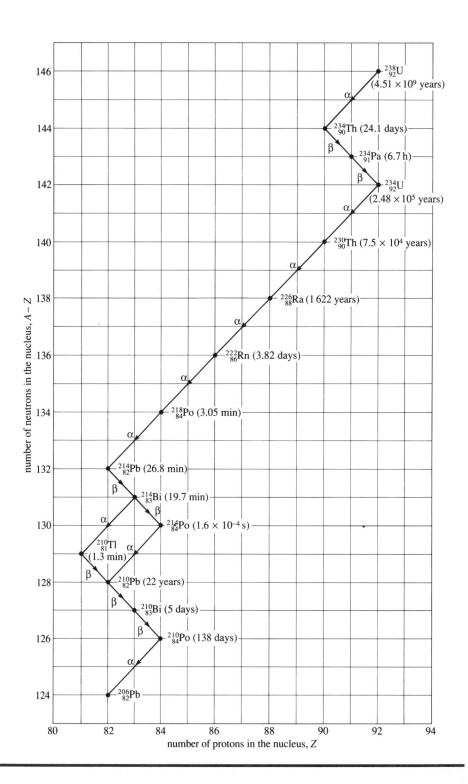

Figure 2.5 The ^{238}U radioactive decay chain. The half-life of each isotope in the chain is given in parentheses. (The concept of *half-life* is discussed in Section 2.2.2.)

▷ From Figure 2.5, what are the stages in the production of radon-222 from uranium-238?

▶ Six radioactive decay processes occur in the formation of radon-222 from uranium-238. An initial α-decay is followed by two β-decays in succession. There are then three more α-decays.

Once radon gas is in the atmosphere it can be breathed in, and while in the lung its radioactive decay can cause biological damage; we shall be discussing this hazard in Chapter 4.

Because the particles and radiation emitted during radioactive decay can be hazardous, the biological hazard arising from a particular radioisotope does not cease until the chain that it gives rise to has reached its end—that is, until a stable nucleus has been formed. We can see this from Figure 2.5. Each member of the chain is radioactive except the last, lead-206. So the radioactive decay of ^{238}U does not mean that the risk has gone away; the risk is now also from the decay of the next and subsequent radioisotopes in the chain. For this reason, the overall risk at any one time is the sum of the contributions from all the radioisotopes in the decay chain. Some radioisotopes are more hazardous than others because they vary in their ease of incorporation into the human body. Hence the relative risk may change at different points in the radioactive decay chain.

2.2.2 The half-lives of radioisotopes

In addition to the *way* in which radioactive atoms decay, the other characteristic of radioisotopes which is very important is *how quickly* they decay. The rate of decay of a radioisotope is characterized by its **half-life**. This is the time that it takes for *half* the atoms in the sample to decay. No matter how many atoms there are, the halving of the number always takes the same time.

Let us now look at radioactive decay slightly more quantitatively. If N_0 is the number of radioactive atoms of a particular isotope at a time $t = 0$, and N_t is the number present at a time t later, then the fraction of the original remaining at time t is (N_t/N_0). The relationship between this fraction and the time that has elapsed is shown in Figure 2.6, where time is measured in units of half-lives, τ, of the radioisotope concerned.

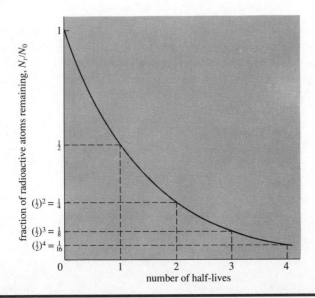

Figure 2.6 Radioactive decay: the relationship between the fraction of the original radioactive atoms remaining and the number of half-lives which have elapsed. Each half-life on the horizontal axis halves the number of atoms of the original radioisotope in the sample.

Let us use the information in Figure 2.6 to find a general relationship between the fraction of radioactive atoms remaining and the half-life, τ. After a time equal to one half-life the fraction remaining, (N_τ/N_0), will be $\frac{1}{2}$, and the number of radioactive atoms remaining, N_τ, will be $(N_0 \times \frac{1}{2})$.

After a time equal to two half-lives the fraction remaining, $(N_{2\tau}/N_0)$, will be $\frac{1}{4}$ ($= \frac{1}{2}^2$), and the number remaining, $N_{2\tau}$, will be $N_0 \times \frac{1}{2}^2$.

After a time equal to three half-lives the fraction remaining, $(N_{3\tau}/N_0)$, will be $\frac{1}{8}$ ($= \frac{1}{2}^3$), and the number remaining, $N_{3\tau}$ will be $N_0 \times \frac{1}{2}^3$.

Can you see the pattern?

$t = \tau \log(N/N_0)/\log(\frac{1}{2})$
p. 2a unit 28 S102

▷ What will be the fraction and number of the radioactive atoms left after a time equal to n half-lives?

▶ After n half-lives the fraction remaining, (N_n/N_0), will be $\frac{1}{2}^n$, and the number of radioactive atoms left will be $N_0 \times \frac{1}{2}^n$.

So, we can write the fraction remaining after n half-lives as

$$N_n/N_0 = \frac{1}{2}^n \qquad (2.7)$$

You can use this equation to find out what fraction of the atoms of any radioactive sample will have undergone radioactive decay, and what fraction remains, after any given period of time.

Question 2.6 ^{235}U and ^{238}U have half-lives of *roughly* 7.5×10^8 and 4.5×10^9 years, respectively (we use rough values to make the calculation easier). Both were present when the Earth was formed (assumed to be 4.5×10^9 years ago). What fraction of each is present in the Earth now? To do this you should first calculate the number of half-lives which have elapsed since the formation of the Earth, and then use Equation 2.7 to calculate what fraction of each remains.

So what can half-lives tell us in general about rates of decay? If we have N atoms of a radioisotope, then, after one half-life, $N/2$ will have decayed. If the half-life is short, those $N/2$ atoms will have disappeared in a short time, and the rates of decay within that half-life must have been large. This is the first important point: a radioisotope will decay quickly if it has a short half-life, and slowly if it has a long half-life.

Now consider what happens during a second half-life. A further $N/4$ atoms will decay—just half of those that disintegrated during the first half-life. Thus, the rates of decay during the second half-life were lower than those during the first. This is the second important point; as time passes and decay proceeds, the rate of decay decreases. We can relate this to Figure 2.6. The rate of decay at any moment is revealed by the steepness of the curve.

▷ When the slope is steep, is the rate of decay large or small?

▶ It is large.

At low values of n, the number of half-lives, the slope is steep; many atoms are disintegrating each second. At high values of n the slope begins to approach the horizontal; much smaller numbers of atoms are disintegrating each second. When the rate of decay is expressed in this way—in terms of the number of disintegrations occurring per second—it is called the **activity** of the sample. The unit of one disintegration per second is known as the **becquerel (Bq)**[*], so activities are expressed in becquerels.

The high activities of radioisotopes with short half-lives, and the decrease of activity with time have important implications for radioactive waste disposal, as you will see in Chapter 5.

Because samples of radioactive atoms will feature so prominently in later discussions, it is very important to note that, unlike chemical reactions, the rate at which they will decay at any particular moment cannot be changed: heating them, subjecting them to high or low pressures, or to any other physical process, does not alter the half-life.

Summary of Section 2.2

1   The important modes of radioactive decay are α-decay, β-decay and positron emission, in which the respective emissions are α-particles ($^4_2$He), electrons (β-particles) and positrons, and the nucleus is transformed into another element. Radiation in the form of γ-rays is usually emitted as well.

2   There are large differences in the amount of material needed to stop these different emissions: α-particles require the least and γ-rays the most.

3   The product of the radioactive decay of an isotope may itself be radioactive, and so on; that is, it might be part of a radioactive decay chain.

4   Radioisotopes decay more quickly if they have short half-lives; as the decay of any radioisotope proceeds, its rate of decay (the activity) decreases. This rate of decay cannot be changed by changing the physical conditions.

2.3   Interactions between neutrons and nuclei

This section is concerned with what happens when nuclei are bombarded with neutrons, such as happens in a nuclear reactor.

We shall start with the particular interaction known as nuclear fission, since it is this process that produces the energy on which nuclear power is based.

2.3.1   Nuclear fission

An introductory aside

Doing scientific research is often rather like a group of people trying to do a complex jigsaw puzzle without having a copy of the final picture. They each assemble separate portions of it until, suddenly, it becomes clear to one of them what the whole picture is. Then, what was a slow, laborious business becomes a mad rush to complete the picture, and to confirm the theory of what it looks like.

The discovery of fission followed this pattern. A number of laboratories throughout the world were doing experiments on the interaction of neutrons with uranium. Each new finding sparked off furious activity, as these laboratories sought both to reproduce the findings of others and to fit together the pieces of the puzzle. It was Otto Hahn and Fritz Strassmann who, in 1938, showed unambiguously that new, lighter, elements were formed when uranium was bombarded with neutrons. In 1939 this finding was

---

[*] The unit is named after Henri Becquerel, who, in 1896, discovered that photographic plates placed near to uranium compounds became fogged, from which he concluded that the uranium was emitting penetrating radiation. This was the first observation of radioactivity.

then explained in nuclear physics terms by Lise Meitner and Otto Frisch, who coined the word *fission* for the process.

It was the outbreak of the Second World War, in 1939, and the entry into it of the United States in 1941, which provided the stimulus to exploit fission as an energy source, initially to make weapons. Both the resources that were deployed and the rate of progress were extraordinary. The first nuclear reactor began working in Chicago, USA, in 1942, and the first nuclear weapon was tested in July 1945.

The fission process and energy production

The early experiments showed that fission occurred when, for example, a nucleus of uranium-235 absorbed a neutron, to produce uranium-236. The uranium-236 nucleus then split into two, to form two **fission products**, and, at the same time, two or three neutrons were emitted. This is called **neutron-induced fission**, to distinguish it from the spontaneous fission mentioned in Section 2.2.1. However, since neutron-induced fission is the subject of this book, we shall simply call this process *fission* throughout.

One of the isotopes which is formed when this fission process occurs is $^{140}_{54}$Xe. Let us use Appendix 1 to answer the following question.

▷ If uranium-236 undergoes fission to produce $^{140}_{54}$Xe and three neutrons, what is the other isotope formed?

▶ The mass number of this xenon isotope is 140. Three neutrons have a mass of 3, so that the mass of the other new isotope is $(236 - 140 - 3)$, which is 93. Xenon has an atomic number of 54 and uranium has an atomic number of 92, so the second new isotope has an atomic number of $(92 - 54)$, that is, 38. Appendix 1 shows that $Z = 38$ corresponds to strontium, so the other isotope formed is $^{93}_{38}$Sr.

We can therefore write the stages in this example of the fission process as

$$^{235}_{92}\text{U} + {}^{1}_{0}\text{n} \longrightarrow {}^{236}_{92}\text{U} \longrightarrow {}^{140}_{54}\text{Xe} + {}^{93}_{38}\text{Sr} + 3\,{}^{1}_{0}\text{n} \tag{2.8}$$

This is just *one* example of the isotopes that can be formed following fission of uranium-235. Another is

$$^{235}_{92}\text{U} + {}^{1}_{0}\text{n} \longrightarrow {}^{236}_{92}\text{U} \longrightarrow {}^{147}_{57}\text{La} + {}^{87}_{35}\text{Br} + 2\,{}^{1}_{0}\text{n} \tag{2.9}$$

Besides those in Equations 2.8 and 2.9, there are other combinations of isotopes which can be produced. In addition, sometimes more than three neutrons will be emitted, and sometimes less; the *average* number of neutrons emitted is about 2.5.

Now that we have looked at an example of fission, we should try to account for two important properties of fission. The first is that the fission products are radioactive, and the second is that energy is produced when it occurs. Could we have deduced this from what we have learnt so far? To enable you to find out, start with the following question.

Question 2.7 Go back to Figure 2.3 and mark on it the positions of uranium-236 and of the two isotopes formed when it undergoes fission according to the example in Equation 2.8. Then answer the following questions about the new isotopes formed:

radioactive

(a) Are they likely to be stable or radioactive?

n

(b) If they are likely to be unstable, is this because, judging by the distribution of stable isotopes in Figure 2.3, they have too many protons or too many neutrons?

β-decay

(c) If they decay, how are they likely to do so?

(d) On the basis of Figure 2.3, give a general reason why the fission products of heavy elements like uranium are likely to be radioactive, and to decay in this way.

We can draw a general conclusion from this example, which is that fission products are normally radioactive, and undergo β-decay. The radioactivity of the fission products is the key to the safety problems associated with nuclear power.

Let us now consider the question of the energy released. To simplify the argument, consider a *hypothetical* example of fission in which no neutrons are released—the case where a neutron is absorbed by $^{235}_{92}U$ and the $^{236}_{92}U$ formed simply splits into two equal halves of mass number 118 and atomic number 46. From Appendix 1, you can see that these halves consist of the palladium isotope $^{118}_{46}Pd$.

▷ From Figure 2.2, is the binding energy per nucleon larger in palladium or in uranium?

▶ It is larger in palladium; the curve slopes downward from $A = 118$ to $A = 236$ and beyond.

The ^{236}U nucleus and the two ^{118}Pd nuclei contain the same number of nucleons. As the binding energy per nucleon is larger in palladium, this means that the two palladium nuclei will also have a greater *total* binding energy than the ^{236}U nucleus, and that fission is accompanied by a release of energy (Figure 2.7).

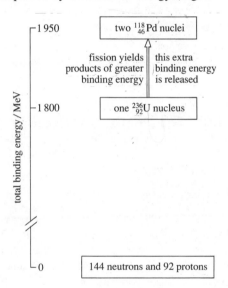

Figure 2.7 A diagram akin to Figure 2.1, which shows the energy change during fission of a ^{236}U nucleus. Because the binding energy per nucleon in palladium is, from Figure 2.2, larger than in uranium, the total binding energy of two $^{118}_{46}Pd$ nuclei is greater than that of one $^{236}_{92}U$ nucleus. Consequently, when fission occurs, this extra binding energy is released, just as in Figure 2.1 the binding energy of the deuterium nucleus was released when it was formed from its constituent nucleons.

Although we have deduced this using the hypothetical fission in Figure 2.7, the principle applies to real examples such as Equations 2.8 and 2.9. When fission occurs, the shape of Figure 2.2 implies that the fission products have larger binding energies per nucleon than the parent nucleus. Consequently, fission is associated with an energy release. Moreover, the energy released per fission is very large: on average it is about 200 MeV. This is nearly 50 million times more energy per atom than the 4 eV obtained by burning carbon. Allowing for the differences in mass of uranium and carbon atoms, this means that, *per unit mass*, fission of uranium produces about 2.5 million times more energy than burning carbon.

The neutron-induced fission of ^{235}U which produces all this energy can occur irrespective of the speed of the bombarding neutron, that is, whether the neutron is moving very quickly or very slowly when it collides with the nucleus. In other words, the isotope can undergo fission with neutrons of *any* kinetic energy. Isotopes that can undergo fission no matter what the kinetic energy of the neutron are said to be **fissile isotopes**. Plutonium-239 and uranium-233 are two other important examples of fissile isotopes. Even neutrons of negligible kinetic energy can cause fission in fissile nuclei. *It is the fissile nuclei that are used to produce fission energy in nuclear reactors.*

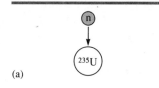

(a)

(b)

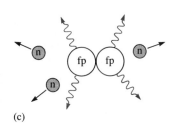

(c)

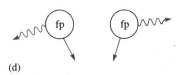

(d)

Figure 2.8 The particles and types of radiation involved in fission: bombardment of a ^{235}U nucleus with a neutron (a) leads to the formation of a ^{236}U nucleus (b), which very quickly undergoes fission (c). Fission products (fp) are formed, and neutrons (n) and γ-rays (wavy arrows) are emitted; subsequently (d), the fission products decay, emitting β-particles (arrows) and γ-rays.

In contrast, there are some nuclei, such as $^{232}_{90}$Th and $^{238}_{92}$U, that will undergo fission providing that the bombarding neutron has a minimum amount of kinetic energy (called the **threshold energy**). The threshold energy for fission is just over 1 MeV for both ^{232}Th and ^{238}U. These nuclei, therefore, are said to be *fissionable*, but non-fissile.

Question 2.8 Say whether the following statements are true or false:

(a) Fission occurs in low-mass elements.

(b) Neutrons with negligible kinetic energy can cause fission in fissile nuclei.

(c) Most fission products are radioactive, and undergo β-particle decay.

The distribution of the energy released in fission

So far we have established that a range of different fission products are produced when fission of a particular isotope occurs, and that a very considerable amount of energy (about 200 MeV) is released. Figure 2.8 shows the process of fission and the products that are formed; Table 2.2 shows how the energy is distributed among the products.

Table 2.2 The distribution of the energy released (about 200 MeV) among the particles and radiation formed by the fission process.

	Fission products	γ-rays, β-particles and antineutrinos	Neutrons
total energy	170 MeV	25 MeV	5 MeV

Most of the energy, typically about 170 MeV, is in the form of kinetic energy of the fission products. The fission-product kinetic energy is then rapidly converted into heat when the fission products interact with other atoms. This heat can then be used to generate electricity, as we shall describe in Chapter 3.

About 25 MeV of the remaining energy is carried by γ-rays, β-particles and antineutrinos.

The neutrons emitted when fission occurs acquire a much smaller amount of energy, about 5 MeV, as kinetic energy. If an average of 2.5 neutrons are emitted per fission, their average kinetic energy is therefore about 2 MeV. Such neutrons are travelling at speeds of about 20 million m s^{-1}. For comparison, Concorde can travel at a speed of 550 m s^{-1}, so neutrons with a kinetic energy of 2 MeV travel about 36 000 times faster than Concorde!

2.3.2 Other interactions of neutrons with nuclei

When a neutron collides with a nucleus, there are two basic types of interaction:

1   the neutron is **scattered** by the target nucleus, or

2   the neutron is **absorbed** by the target nucleus.

Fission is just one of the events that may be the outcome of the *second* of these two kinds of interaction; that is, it comes under the heading of absorption. We now consider the other possibilities—scattering, and absorption events other than fission.

Scattering of neutrons

By scattering we mean that a neutron moving in a certain direction collides with the nucleus and, as a result, moves off in a different direction. How does this affect the neutron's kinetic energy? To find out, you should now do Activity 2.1.

Activity 2.1

In this activity, you will use collisions between familiar objects as a *model* of neutron scattering in a nuclear reactor. First study Figure 2.9, reading the caption carefully. We hope that it conforms with your own experience! Now answer the following questions:

(a) Two fission-generated neutrons have an energy of 2 MeV and are moving at about 20 million m s^{-1}. One of the neutrons finds itself in an environment of hydrogen nuclei, ^{1}H, with which it undergoes a number of successive collisions. The other neutron undergoes the same number of successive collisions with uranium-238 nuclei, ^{238}U. In which of the two cases will the neutron have the lower speed and kinetic energy after the collisions?

(b) Suppose that a neutron is scattered in collisions with each of the following types of nuclei: carbon, deuterium ($^2_1$H), gold, magnesium, oxygen, sodium and uranium-238. Use Appendix 1 to list the nuclei in order of the average decrease in neutron energy which occurs due to the collision.

(c) Models are often introduced for a special purpose, and may not, therefore, be able to account for *all* aspects of the behaviour they are applied to. In this case, our marble/cannon-ball model has been used to tell us something about the way neutrons exchange energy with nuclei during scattering. Name one of the outcomes of neutron–nucleus encounters that you have met which the model cannot reproduce.

BEFORE COLLISION   AFTER COLLISION

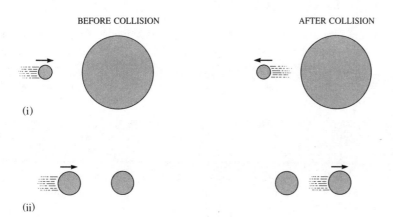

(i)

(ii)

Figure 2.9   In (i), a marble undergoes a head-on collision with a stationary cannon-ball. The cannon-ball hardly moves, and the marble rebounds with a speed that is little different from that with which it entered the collision: it has lost very little of its kinetic energy. In (ii), the collision is with a second marble. The stationary marble moves off sharply but the incoming marble is reduced almost to a standstill: nearly all of its kinetic energy has been transferred to the outgoing marble. In glancing, as opposed to head-on collisions, the principle still holds: the moving body loses much more kinetic energy if it strikes a body of similar mass, than if it strikes a much heavier object. To sum up, the *average* energy lost by a moving body in a scattering collision decreases as the mass of the body it strikes increases.

To summarize, fast-moving neutrons in a nuclear reactor lose energy much more quickly if they are scattered in collisions with light nuclei rather than with heavy ones. This principle has important consequences for some features of nuclear reactor design.

Absorption of neutrons

The other possible outcome of an encounter between a neutron and a nucleus is neutron absorption. This may have one of three consequences. Firstly, the neutron and the nucleus may simply coalesce: this is called **radiative capture**. We discussed one example, the absorption of a neutron by a hydrogen nucleus, ^{1}H, in Section 2.1.2 (Equation 2.2). *Some* of the collisions between neutrons and ^{238}U nuclei, which is the most common isotope of uranium, have this sort of outcome:

$$^{238}_{92}\text{U} + ^1_0\text{n} \longrightarrow ^{239}_{92}\text{U} \tag{2.10}$$

Secondly, for some elements, the absorption of the neutron may lead to the ejection of a charged particle such as a proton or an α-particle. An example of **charged-particle emission** of great importance in the nuclear industry is neutron absorption by boron-10 nuclei:

$$^{10}_5\text{B} + ^1_0\text{n} \longrightarrow ^7_3\text{Li} + ^4_2\text{He} \tag{2.11}$$

Finally, as we have seen in Section 2.3.1, neutron absorption may, in the heavier elements, lead to nuclear fission:

$$^{235}_{92}U + {}^{1}_{0}n \longrightarrow {}^{140}_{54}Xe + {}^{93}_{38}Sr + 3{}^{1}_{0}n \qquad (2.8)$$

The different ways in which neutrons can interact with nuclei are summarized in Table 2.3. Notice that γ-rays are emitted in all three cases of neutron absorption.

Table 2.3   The different neutron interactions with nuclei.

Basic interaction	The outcome
Scattering	Neutron loses kinetic energy and the target nucleus gains kinetic energy; the average energy loss of a neutron decreases as the mass of the target nucleus increases.
Absorption	
radiative capture	Neutron absorbed by the target nucleus and only γ-rays are emitted.
charged particle emission	Neutron absorbed by the target nucleus and a charged particle is emitted together with some γ-rays.
fission	Neutron absorbed by the target nucleus to form a new nucleus, which splits into two different fission products; neutrons are emitted together with some γ-rays.

Summary of Section 2.3

1   Neutron-induced fission leads to the splitting of a nucleus of high mass number, such as ^{235}U, into two fission products and, usually, 2 or 3 neutrons. The fission products are radioactive and generally undergo β-decay.

2   In nuclear fission, about 50 million times more energy is released per atom than in the burning of carbon.

3   When the fission of a nucleus can be induced by neutrons of any energy, the nucleus is said to be *fissile*. Fissile nuclei include ^{235}U, ^{239}Pu and ^{233}U, and are used as fuel in nuclear reactors.

4   Fission is a particular outcome of the absorption of a neutron by a nucleus. Other possible outcomes of neutron absorption are radiative capture and charged particle emission.

5   Instead of being absorbed by a nucleus, a neutron may simply be scattered by it. The lower the mass of the scattering nucleus, the larger is the average kinetic energy loss experienced by a neutron on scattering.

2.4 The dependence of nuclear reaction probabilities in uranium on neutron energy

You have just looked at the different reactions that neutrons can undergo when they interact with a nucleus. This section is concerned with studying how *probable* these different reactions are. The reason why this is important is that you will then be able to understand many of the features of nuclear reactor design more clearly.

Suppose that we want to bring about fission in uranium. We can do this by bombarding uranium with neutrons. But the last section pointed to reasons why this process might be inefficient: for example, instead of causing fission, the neutrons may be scattered by the uranium nuclei, or they may undergo radiative capture. What steps can we take to increase the likelihood that the neutrons will cause fission?

One property that bears crucially on this question is the energy of the bombarding neutrons. Let us first consider those neutrons with an energy of 1 MeV; they are described as **fast neutrons**, and are moving at about 14 million m s^{-1}, a speed and energy similar to that of some of the neutrons produced in the fission of ^{235}U (Section 2.3.1). Suppose that neutrons with this energy are let loose in a sample of natural uranium. What will be the consequence?

Nuclear physicists answer such questions by considering the differing probabilities of the various events that may occur when a neutron enters a uranium sample. One possible outcome is the desired one of the fission of a nucleus of the fissile isotope ^{235}U. However, in *natural* uranium ^{235}U is far from being the most abundant nucleus; that role is fulfilled by ^{238}U. Natural uranium contains 99.28% ^{238}U and only 0.72% ^{235}U: of every 1 000 uranium nuclei, only 7 are ^{235}U. Considering all the possible interactions, it turns out that if our neutron has an energy of 1 MeV, scattering of that neutron by a ^{238}U nucleus is by far the most likely consequence. However, for the chain reaction that will be discussed in Section 2.5.2 to be possible, fission needs to take place in the fissile ^{235}U component.

▷ How do you think that it might be possible to increase the probability that 1 MeV neutrons will have a fission interaction with ^{235}U in a sample of uranium?

▶ If we make uranium that contains a much greater proportion of ^{235}U than in natural uranium, then there will be a much greater chance that fast neutrons will encounter ^{235}U nuclei and cause fission rather than being scattered by ^{238}U nuclei.

Uranium whose ^{235}U content has been artificially increased above the natural level is said to be **enriched uranium**. From what we have said earlier in this section, you should realize that if we want to use fast neutrons to bring about fission in uranium we should use uranium highly enriched in the fissile isotope ^{235}U.

Now let us turn to neutrons with a somewhat lower energy. Until now, there has been no need to consider the possibility of radiative capture by ^{238}U, because with 1 MeV neutrons it is insignificant. However, there is a neutron energy range between 10 eV and 1 000 eV (1 keV) where radiative capture by ^{238}U (Equation 2.10) may become as much as 10 000 times more likely than at 1 MeV. This is known as the **resonance absorption region**. Neutrons in this energy range are very likely to be absorbed by ^{238}U nuclei and hence become unavailable for ^{235}U fission.

Finally we turn to neutrons with even lower energies. A suitable instance is that of a stray neutron which has spent some time in the air around us. This would have an energy of about 0.025 eV and be moving at about 2 200 m s^{-1}. Neutrons of this kind are called **thermal neutrons**, because they are in thermal equilibrium with their surroundings. In this case, the situation is very different from the one that prevails with neutrons in the resonance absorption region. In particular, at these low energies, it turns out that the probability of a neutron inducing fission in ^{235}U is greatly increased at the expense of other types of encounter between neutrons and uranium nuclei. For example, if the uranium sample contains equal numbers of ^{235}U and ^{238}U nuclei, the chance that a neutron will cause ^{235}U fission is over 200 times the chance that it will be captured by ^{238}U. This ratio is even greater than the 140 : 1 preponderance of ^{238}U in natural uranium. Consequently, even in natural uranium, thermal neutrons are more likely to induce fission than to be captured. Even with thermal neutrons, however, enrichment of the uranium in ^{235}U will improve the chances that a neutron will induce fission.

You now know enough about the way in which neutrons interact for us to consider how a nuclear reactor works.

Summary of Section 2.4

1 The chance that a neutron will cause fission in ^{235}U is especially large for thermal neutrons, whose energies are low (less than 1 eV).

2 The chance that a neutron will undergo a radiative capture reaction with ^{238}U is greatest at neutron energies between 10 eV and 1 keV, an energy range called the *resonance absorption region*.

3 Fast neutrons (with energies of around 1 MeV) in natural uranium are far more likely to be scattered by ^{238}U nuclei than to induce fission in ^{235}U. This is mainly because the ^{238}U nuclei are much more abundant than ^{235}U in natural uranium. Consequently, in order to induce fission in uranium with fast neutrons, it is very important that the uranium should be enriched in ^{235}U above the natural abundance of 0.72%.

2.5 The chain reaction and criticality

In this section we shall put together many of the ideas discussed so far in this chapter to show how nuclear fission can be made to yield a continuous supply of energy, as a product of a chain reaction. We shall also examine how such a chain reaction can be controlled.

2.5.1 Introduction

Let us start by revising some ideas that will be important in this section.

▷ What nuclear particles are emitted when fission occurs, and what is their average kinetic energy (Section 2.3.1)?

▶ Two or three neutrons are emitted when fission occurs, and their average kinetic energy is about 2 MeV.

▷ What type of element (high or low mass number) is required to give the largest average neutron kinetic energy losses in a scattering reaction (Section 2.3.2)?

▶ In answering Activity 2.1 in Section 2.3.1, you showed that low mass number elements gave the largest average neutron kinetic energy loss following a scattering reaction.

Important examples of these low mass number elements are hydrogen (both 'normal' hydrogen, $^{1}_{1}$H, and deuterium, $^{2}_{1}$H) and carbon. Using this information, together with what you learned in the previous section, we can now discuss the fission chain reaction.

2.5.2 The chain reaction

Two points make nuclear fission of uranium-235 a valuable energy source. Firstly, a large amount of energy (about 200 MeV) is released in each fission. Secondly, two or three neutrons are emitted when a fission occurs. These neutrons make it possible to obtain a continuous supply of energy by means of what is called a **chain reaction**. Figure 2.10 shows how this works.

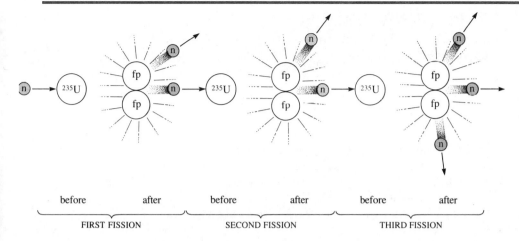

Figure 2.10 Nuclear fission taking place in a chain reaction. Three fissions are shown, and in each fission, two fission products (fp), and either 2 or 3 neutrons (n) are produced. If things can be arranged, as they are here, so that one neutron from each fission successfully initiates a subsequent fission, then the fissions are linked in a chain; that is, there is a chain reaction. Energy can then be extracted continuously from a nuclear reactor while such chain reactions proceed within it.

How can the phenomenon in Figure 2.10 be made to occur in practice? First of all, fissile material is needed. The most widely used nuclear fuel is uranium, in which the fissile isotope is ^{235}U; we mentioned in Section 2.4 that uranium-235 forms only 0.72% of *natural* uranium, but the percentage can be increased if necessary (as you will see in Chapter 5). The uranium is made up into rods; a nuclear reactor will have several hundred—and sometimes thousands—of these, arranged to form the reactor's **core**. Given this arrangement, there are then two ways in which a chain reaction can be made to occur in practice.

The first way is the one suggested by point 3 of the summary of Section 2.4: the fuel can contain a large percentage (20–30%) of a fissile isotope such as ^{235}U or ^{239}Pu. This high percentage means that the fast neutrons produced by the break-up of one fissile nucleus have an excellent chance of encountering and inducing fission in another, before scattering lowers their energy into the resonance absorption region, where they would be swallowed up by ^{238}U. In this arrangement, fission is induced by fast neutrons, so a reactor that exploits it is called a **fast reactor**.

The second way that a chain reaction can be made to occur exploits point 1 of the Section 2.4 summary, and what you deduced in Activity 2.1. This time, the fuel rods contain a much lower proportion of the fissile isotope (0.72–10%), but in this instance the fuel rods are surrounded by a material containing nuclei of low mass such as hydrogen, deuterium or carbon (graphite). This material is called a **moderator** (Figure 2.11). Again, fission produces fast neutrons with energies of about 1–2 MeV. This time, however, they are much less likely to induce fission in other nuclei because the percentage of fissile nuclei is much smaller. Many of the neutrons therefore leave the fuel and enter the moderator.

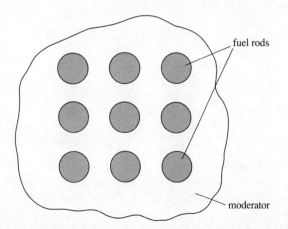

Figure 2.11 Section of the core of a thermal nuclear reactor, showing the arrangement of the fuel rods and moderator. Later, we shall see how this idealized picture is modified by the need for a coolant to remove heat from the fuel rods.

▷   What happens to the energy of fast neutrons in the moderator?

▶   Scattering by the light nuclei will quickly lower their energies to thermal values (around 0.025 eV): they become thermal neutrons.

If, with these low energies, they then re-enter a fuel rod, their chances of inducing fission in a fissile nucleus are now very much greater (point 1 of the Section 2.4 summary).

It is especially important that the reactor design ensures that the lowering of the neutron energy from 1–2 MeV to 0.025 eV takes place *in the moderator*. This allows the energy to drop through the resonance absorption region when the neutron is out of contact with ^{238}U nuclei, and so greatly reduces the chance of neutrons being captured by ^{238}U. The neutrons are then available for continuing the chain reaction with ^{235}U nuclei. These points are summarized in Figure 2.12. A reactor that includes a moderator and exploits these principles is called a **thermal reactor** because in it, fission is induced primarily by thermal neutrons.

Figure 2.12 The significance of different neutron energies in a thermal reactor. The division into energy regions is idealized to illustrate the principle. Note that the energy scale is logarithmic: successive powers of ten occur at even intervals along the horizontal axis. On the right of the energy scale, neutrons with energies of around 10^6 eV (1 MeV) or more are produced in the fuel by fission of ^{235}U nuclei. They leave the fuel and then spend time in the moderator as their energies are reduced from about 10^6 eV to below 1 eV. This phase covers the middle region of the scale, and includes the resonance absorption region in ^{238}U. The neutrons finally pass into the left-hand region where, if they leave the moderator and re-enter the fuel, they have an excellent chance of causing fission in other ^{235}U nuclei.

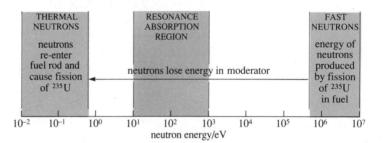

The picture we have just drawn, of the chain reaction in the core of a thermal reactor is, of course, an idealized one. In practice, *some* neutrons will be captured by ^{238}U nuclei or cause fission in ^{235}U nuclei before they leave the fuel, and others will leave the fuel but only lose a little energy in the moderator before encountering a fuel rod again, when they may be absorbed by ^{238}U. Hence, you must assume that the ideal behaviour we have outlined will apply to the *majority* of the neutrons rather than to *all* of them.

The role of the reactor designer is to make as many neutrons as possible conform to the ideal case, by careful choice of:

(i)   the amount of fissile material in the fuel;

(ii) the fuel rod diameter;

(iii) the fuel rod spacing within the moderator;

(iv) the choice of moderator used to slow the neutrons down.

2.5.3 Critical size and critical fuel mass

For a chain reaction to be a useful source of energy, the fissions have to occur at a steady, and controlled, rate. If the rate is constant, then the chain is said to be *self-sustaining*; that is, just one of the neutrons from each fission in the chain goes on to cause another fission (see Figure 2.10). This is called a **critical system**.

Apart from the type of fuel and moderator, the main factor determining whether a system is critical is its *size*. Neutrons reaching the surface of a reactor core can leave it entirely: they 'leak out', and cannot take part in the chain reaction. This can occur with reactors of any size, but the *proportion* of neutrons leaking out increases as the core gets smaller. The minimum size (that is, dimensions) at which a critical system becomes possible for a particular type and arrangement of fuel is called the *critical size*.

We have said that a critical system is one in which the chain reaction is self-sustaining. If we make a reactor that is *bigger* than its critical size, then a smaller proportion of the neutrons escape, and more become available to cause fission. We thus have the situation in which more than one of the neutrons produced in each fission goes on to cause another fission. In consequence, the number of fissions and the number of neutrons in the reactor core increase steadily (Figure 2.13); this is called a **supercritical system***. Conversely, if we make our system smaller than the critical size, the chain reaction stops; that is, the chain is broken. This is called a **subcritical system**.

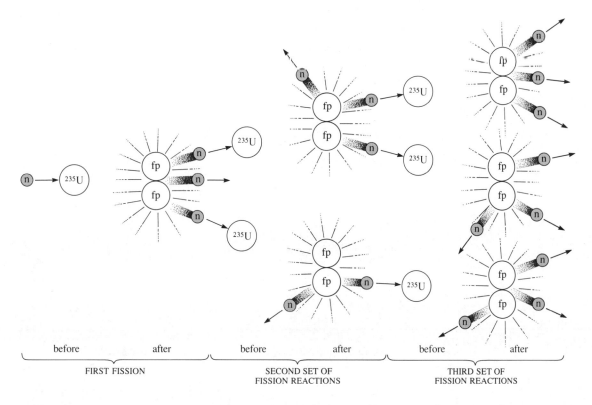

before   after   before   after   before   after

FIRST FISSION   SECOND SET OF
FISSION REACTIONS   THIRD SET OF
FISSION REACTIONS

We now ask what would happen if a neutron-absorbing material such as boron (Equation 2.11) were put into a critical system. The number of fissions taking place would be reduced, because there would be an increase in the number of fission neutrons being absorbed. The system would therefore become subcritical. Conversely, if a neutron absorber is removed from a critical system there will be a decrease in the number of neutrons being absorbed, and the system will become supercritical. These principles are exploited whenever neutron-absorbing control rods are used to lower or raise the power level in a nuclear reactor (Section 3.1.6).

Finally, we must correct a misconception that might have arisen.

We have spoken so far as if there is only *one* chain reaction occurring in a critical system. In fact, in a typical nuclear power station producing 1 000 MW of electricity, there are some 10^{19} fission chains being propagated at any one time. Clearly, the idea that there is a single chain reaction is totally wrong!

* This is the situation that is required when constructing a nuclear weapon, as you will see in Chapter 8.

Figure 2.13 The chain reaction in a supercritical system. The first fission produces three neutrons, two of which produce the two fissions in the second set. Of the five neutrons produced by the two fissions in the second set, three go on to cause the three fissions in the third set. The essential feature of a supercritical system is that the number of fissions per second and the number of neutrons increase as the chain reactions proceed. Contrast this with the critical system in Figure 2.10, in which there is just one fission at each stage, and the number of fissions per second is constant.

Summary of Section 2.5

1 A self-sustaining chain reaction can be established, in which the fission rate (the number of fissions per second) is constant. This is called a *critical system*.

2 A *subcritical system* is one in which the number of fissions per second is decreasing; a *supercritical system* is one in which the number of fissions per second is increasing.

3 Critical systems can be made based on neutrons with high (around 1 MeV) and low (below 1 eV) energy. Reactors that operate using high-energy neutrons (fast neutrons) are called fast reactors; those that use low-energy neutrons (thermal neutrons) are called thermal reactors.

4 In thermal reactors, scattering of the neutrons produced in fission reactions by low-mass moderator nuclei slows the neutrons down before they cause fission.

5 The chain reaction can be controlled by changing the amount of neutron-absorbing material in the reactor core.

2.6 The production of fissile isotopes

One of the limitations of nuclear power which we shall examine later is posed by the size of world reserves of uranium, and hence by the availability of ^{235}U as the fissile fuel. Therefore, if ^{235}U were the only fissile isotope used as a nuclear fuel, uranium could not make a significant contribution to the world's future energy supplies. There is, however, a way round this problem.

So far we have mentioned that a neutron can interact with ^{238}U and undergo radiative capture, thus preventing that neutron from participating in the chain reaction. But this radiative capture can be used to advantage, because it leads to the creation of another fissile isotope.

▷ By looking back at Table 2.3 and Equation 2.2, remind yourself what radiative capture of neutrons is, and then say what isotope is formed when a neutron undergoes a radiative capture reaction with a ^{238}U nucleus.

▶ In radiative capture a neutron is absorbed, but only γ-rays are emitted. The number of protons in the nucleus (92 for uranium) is therefore unchanged. The number of neutrons will, however, be increased by one. The mass number of the new isotope will therefore be (238 + 1), and the new isotope will be $^{239}_{92}$U. In fact this reaction was used as an example of radiative capture in Equation 2.10.

The isotope ^{239}U, however, is unstable: it undergoes β-decay (half-life 23.5 minutes) to form neptunium-239 (^{239}Np), which is also unstable, itself undergoing β-decay (half-life 2.36 days) to form ^{239}Pu. Even ^{239}Pu is unstable (undergoing α-decay to ^{235}U), but its half-life is so long (24 000 years) that it can be considered as stable for all practical purposes. We can write this sequence of nuclear reactions as:

$$\ce{^{238}_{92}U} + \ce{^{1}_{0}n} \xrightarrow[\text{capture}]{\text{radiative}} \ce{^{239}_{92}U} \xrightarrow{\beta\text{-decay}} \ce{^{239}_{93}Np} \xrightarrow{\beta\text{-decay}} \ce{^{239}_{94}Pu} \tag{2.12}$$

Now we have already mentioned (Section 2.3.1) that ^{239}Pu is a fissile isotope, like ^{235}U, whereas ^{238}U is only fissionable. So what has been done is to transform something that only undergoes fission at high neutron energies (^{238}U) into something that undergoes it at all neutron energies. You will discover another example in Activity 2.2. Do this now.

Activity 2.2

(a) Natural uranium contains 99.28% ^{238}U and only 0.72% of the fissile isotope ^{235}U. But you have just seen how, if the predominant but non-fissile isotope, ^{238}U, is bombarded with neutrons, it is transformed in three stages into the fissile isotope, $^{239}_{94}$Pu. In principle, to roughly what extent does this process increase the amount of fissile material to which we have access?

(b) The only other relatively common element whose atomic number is close to that of uranium is thorium. It is about four times as abundant in the Earth's crust as uranium, and consists entirely of the non-fissile isotope $^{232}_{90}$Th. When $^{232}_{90}$Th is bombarded with neutrons, it undergoes a three-step transformation exactly analogous to that of $^{238}_{92}$U. By using Appendix 1, write an equation for each of the three stages. Examine Section 2.3.1 and you will find out if the end product is fissile or non-fissile. If this process were to be carried out along with the transformation of $^{238}_{92}$U, to what extent might the amount of fissile material now be increased?

Both the examples of the production of a fissile isotope which we have given involve absorption of a neutron. Since the fission reaction results in the emission of neutrons, it follows that putting either ^{238}U or ^{232}Th into a reactor and allowing them to absorb some of the neutrons will result in the production of new fissile fuel. Obviously, the reactor would have to be designed to allow for this, so that this extra neutron absorption does not stop the chain reaction.

Rather confusingly, the production of new fissile fuel in this way is given different names depending on how much *new* fissile material is produced relative to the amount of the original fissile fuel which has been used up. If the amount of new fissile material produced is *less* than the amount of the original fissile material used to produce it, then the process is called **conversion**, a process that occurs in thermal reactors. If the amount produced is *more* than the amount of original fuel used, then it is called **breeding**. ^{238}U and ^{232}Th are called **fertile isotopes** because they can be transformed into fissile material.

Summary of Section 2.6

1 If the only available fissile isotope were naturally occurring ^{235}U, which comprises just 0.72% of uranium, the potential of nuclear fission as an energy source would be very limited.

2 In a nuclear reactor, the more common isotope ^{238}U (99.28% of uranium) can undergo neutron absorption followed by two β-decays to form the fissile ^{239}Pu. Similarly, the even more common ^{232}Th will yield the fissile ^{233}U. These processes could, in principle, bring about a several hundredfold increase in the supply of fissile material.

3 When the reactor in which such reactions occur produces more new fissile material than it consumes, the process is called *breeding*; if it produces less, the process is called *conversion*.

Now test your mastery of Chapter 2 by trying Question 2.9.

Question 2.9 Choose the correct italicized word or expression to complete the following statements:

(a) The energy released in chemical reactions is normally much (*greater/less*) than in nuclear ones.

(b) Nuclear binding energy is released when a neutron is absorbed by a nucleus. This energy arises because the mass of the new nucleus is (*greater/less*) than that of the original nucleus and neutron separately.

(c) Stable nuclei are formed by (*all/only a limited number*) of combinations of the mass number, A, and atomic number, Z.

(d) β-decay is accompanied by the emission of (*an electron/a positron*).

(e) Fissile isotopes are nuclei that undergo fission when they interact with (*high-energy neutrons only/neutrons of any energy*).

(f) Fission products are radioactive, and normally decay by (*β-emission/α-emission*).

(g) (*Most/very little*) of the energy released in fission is given to the fission products as kinetic energy. This kinetic energy is very rapidly turned into heat energy.

(h) The average energy lost by a neutron when it is scattered by a nucleus is (*more/less*) for high-mass nuclei than for the low-mass ones.

(i) Radiative capture of a neutron by a nucleus results in (*emission of γ-rays only/emission of γ-rays together with a charged particle*).

(j) Whether fission, radiative capture or neutron scattering occurs when a neutron encounters a uranium nucleus depends (*very little/very strongly*) on the isotope concerned and on the neutron energy.

(k) A critical system is one in which the number of fissions is (*increasing/constant*) with time.

(l) A critical system can be made subcritical by (*adding/removing*) a neutron absorber.

(m) A neutron is most likely to cause fission in ^{235}U at neutron energies (*above/below*) 1 eV, and this is exploited in thermal reactors.

(n) Radiative capture of a neutron by ^{238}U, followed by two β-decays, results in the production of the fissile isotope (*$^{233}U/^{239}Pu$*).

3 Nuclear power: the technology

In this chapter we are going to see how the principles that were discussed in Chapter 2 determine the design of a nuclear power reactor. However, before we consider how a *nuclear* power station works, we shall look at how 'conventional', fossil-fuelled, power stations operate, because many of the features are common to both. Figure 3.1 illustrates these features.

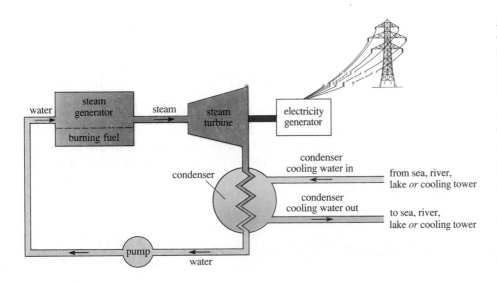

Figure 3.1 The main components of a typical fossil-fuelled power station: the burning of a fossil fuel, such as coal, heats water in the steam generator, and converts it to high-temperature steam, which then passes through a turbine, rotating the turbine shaft. This motion is used to generate electricity (as in a bicycle dynamo). After it has passed through the turbine, the steam is liquefied in a condenser, which is continuously cooled by water obtained from a natural source or from cooling towers. The liquefied water is then returned to the steam generator.

▷ How does a nuclear power station differ from a fossil-fuelled one?

▶ The *main* difference is that the burning of fossil fuel is replaced by the fission of uranium (or plutonium) as the source of heat.

Factors affecting the efficiency of electricity production

The efficiency with which power stations convert their source of heat (fossil fuel *or* nuclear fuel) into electricity, is defined as

$$\text{efficiency} = \frac{\text{electrical power out}}{\text{heat produced}} \times 100\% \qquad (3.1)$$

In general, power station efficiency is between 30 and 40%. Where does the other 60 to 70% of the energy go to? The answer is that, if the power station produces *only* electricity, the rest of the energy is *wasted*. The energy loss occurs because the turbine cannot extract all the energy from the steam. The unextracted energy is transferred as heat to the cooling water of the condenser when the steam is liquefied, and is then carried away[*].

The principal way in which the efficiency of a power station producing only electricity can be maximized is to have the temperature of the steam going into the turbine as high as possible. The steam from fossil-fuelled power stations is at about

---

[*] In so-called 'combined heat and power stations', some of the waste heat is used to heat buildings. This increases the total percentage of heat used usefully, but decreases the amount of electricity produced.

600 °C, which is as high as is practicable; at this temperature the efficiency of the power station is about 40%. If the steam temperature is 350 °C, then the efficiency drops to about 30%.

3.1 The components of a nuclear power station

In this section we shall describe in turn the main components of a nuclear power reactor, drawing our examples mainly from thermal reactors, which are still by far the most common.

3.1.1 Nuclear fuel

At the heart of a nuclear reactor power station is the reactor core. In the core is the nuclear fuel, the material containing the fissile isotope, which, by undergoing neutron-induced fission, provides the energy that ultimately produces electricity in an arrangement like that in Figure 3.1. The fissile isotope is usually ^{235}U, but sometimes is ^{239}Pu. In either case it is mixed with ^{238}U.

The fuel is arranged in the form of a cylindrical rod (or a stack of cylindrical pellets) in a thin-walled metal container, the containing metal being known as **cladding**. The encapsulated rods or pellets and their cladding, which vary in diameter from 5–25 mm are called **fuel pins**, or **fuel rods**. In most reactors, a **fuel element** will consist of a bundle of fuel pins, with spaces between the pins for coolant to flow (Figure 3.2), and the core of the reactor will have two hundred or more fuel elements in it.

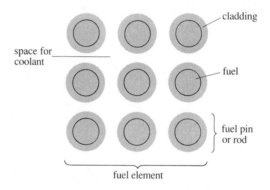

Figure 3.2 Cross-section of a nuclear fuel element. The frame that holds the fuel pins together to make a fuel element is not shown. In practice, each fuel element contains many more fuel pins than the nine shown here.

The fuel itself is normally in the form of the dioxide of uranium or plutonium (UO_2 or PuO_2), but the percentage of fissile isotope in the fuel varies. Natural uranium can only be used if other materials in the core have a low probability of absorbing neutrons, because, if neutrons are lost by absorption in non-fissile materials in the core, the loss may be enough to prevent a self-sustaining chain reaction from being established. In the language of Section 2.5.2, the reactor system cannot be made critical.

This problem becomes significant in a uranium-fuelled reactor that uses *light* water (^1H_2O) for moderating the neutrons and/or cooling the core. As you know (Section 2.5.2), moderators are low-mass nuclei, and either light water or *heavy* water (D_2O) are suitable. However, although it is cheap, the *disadvantage* of using H_2O (as opposed to D_2O) is that ^1H has a significant tendency to *absorb* neutrons. Consequently, in a reactor cooled or moderated by light water, the uranium must be enriched in ^{235}U in order to obtain a critical system (Section 2.5.3). Typical enrichment values are between 1.5 and 4.0%, compared to 0.72% ^{235}U in natural uranium, although, as we shall see, some reactors use fuel with a much higher enrichment than this.

▷ Why does enriching the fuel in ^{235}U increase the chance of fission occurring?

▶ Because it increases the concentration of ^{235}U nuclei in the core, and there is then a greater chance that a neutron will encounter a fissile nucleus (Section 2.4).

We have seen earlier that the majority of the fission products are radioactive, and we obviously want to prevent them from escaping from the reactor. The simplest way to do this is to keep them *in the fuel*. This is the purpose of the containing metal or cladding (Figure 3.2). The choice of cladding material is important, because it should not absorb too many neutrons. However, other factors, for example the need to resist chemical attack by the coolant, may mean that a compromise is necessary.

In a 'conventional', fossil-fuelled, power station the waste ash has to be removed, and new fuel supplied. A nuclear power station is similar: new fuel has to be supplied, and the **spent fuel**, with its radioactive fission products, removed. However, because the amount of energy per gram of fuel obtained from fission is about 2.5 million times greater than that from chemical combustion (Section 2.3.1), the amount of fuel required is far smaller.

Because it is so efficient at producing energy, nuclear fuel produces very little waste. For example, a reactor producing 1 000 MW of electricity with an efficiency of 32% uses only about 1.1 t of fissile material per year; the total mass of fuel (including the non-fissile ^{238}U) varies according to the type of reactor, but is about 100 tonnes. As fission products are the waste material of nuclear fission, the annual mass of the fission product waste will differ little from the 1.1 tonnes of fissile material that is consumed.

In practice, a nuclear fuel element will produce energy for 3–6 *years* before it needs to be replaced.

3.1.2 Coolant

The **coolant** is a gas or liquid that passes through the hot reactor core and carries heat away to the steam generators, where steam is produced. It passes up the outside of the fuel pins (Figure 3.2). The coolant is therefore in contact with the cladding, rather than the fissile fuel. Apart from being able to remove heat from the fuel pins efficiently, coolants should not react chemically with the fuel cladding, or with any other part of the cooling circuit, and they should not absorb too many neutrons.

▷ Why is it undesirable for the coolant to absorb neutrons?

▶ Absorption of neutrons prevents them from participating in the chain reaction. The fuel might then need to be further enriched to compensate for this absorption (and enrichment is expensive).

The coolant should also not be too expensive. Taking all these factors into account, the coolants used in the majority of reactors are the gases carbon dioxide and helium, and light and heavy water. (Only the fast reactors, which will be discussed in Section 3.4, use anything different.)

If a gas coolant is used, then the heated gas is pumped to the steam generator, where steam is produced. If ordinary light water is used, it may be allowed to boil, and hence to produce steam directly for use in the turbine: this is called a **direct steam cycle** (Figure 3.3a). Alternatively, the light water may be prevented from boiling, and the hot water pumped to a steam generator, where steam is produced by heating a *separate* water supply: this is called an **indirect steam cycle** (Figure 3.3b).

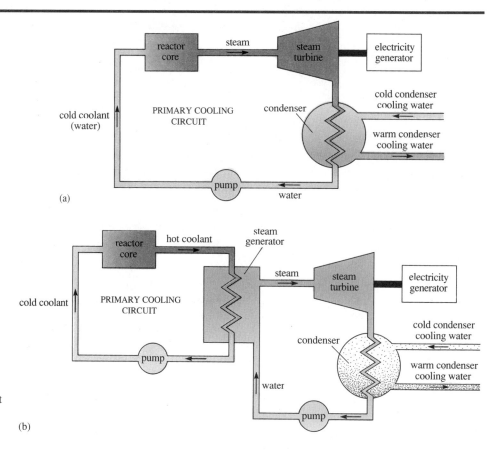

Figure 3.3 Direct (a) and indirect (b) steam cycles in nuclear power stations.

One disadvantage of a direct cycle is that a tiny fraction of the coolant inevitably leaks from the turbine. Light water is so cheap that from an economic point of view this does not matter, but heavy water is extremely expensive. For this reason, when a heavy-water coolant is used, an indirect steam cycle is chosen.

In all cases, the coolant is pumped back to the reactor after it has given up its heat. The path taken by the coolant through the reactor, into the steam generator or turbine and back to the reactor is called the **primary cooling circuit** (or *cooling loop*). Each reactor has several primary cooling circuits, so that if one fails, the necessary cooling is provided by the others.

Pressurized coolants

Let us briefly look at the significance of having light or heavy water as a reactor coolant, either in a direct or an indirect steam cycle. Remember that the efficiency with which a turbine extracts energy from steam depends on the steam temperature. Now at normal, atmospheric, pressure the boiling temperature of water is 100 °C. Earlier, we said that a power station working with steam at 350 °C would have an efficiency of only 30%. You can see that the efficiency would be *very* low indeed if steam at 100 °C were used.

To overcome this problem, the fact that the boiling temperature of water depends on pressure is exploited: if the pressure is raised, the boiling temperature is increased. Indeed, the boiling temperature of normal (light) water increases from 100 °C up to a maximum of 374 °C, by increasing the pressure from 1 atmosphere to 217.7 atm (Figure 3.4); above 374 °C water cannot be kept in the liquid state.

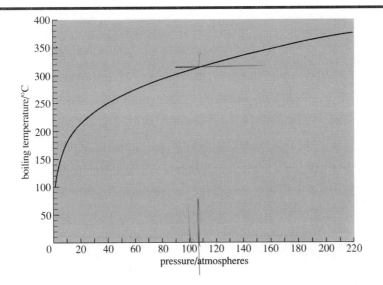

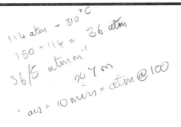

Figure 3.4 Variation of boiling temperature of light water with pressure.

When steam is produced in the steam generators, its temperature will depend on the temperature of the hot coolant that does the heating. So, to increase the efficiency with which energy is extracted in the turbine, the coolant temperature needs to be as high as possible. Liquid water can only reach satisfactorily high temperatures if it is **pressurized**. Pressurization therefore allows the water to be used at a higher temperature, and hence enables the electricity to be generated more efficiently.

▷ Using Figure 3.4, how can we raise the boiling temperature of light-water coolants in a reactor from 100 to 300 °C?

▶ The boiling temperature of light water can be raised from 100 to 300 °C by increasing the pressure to about 87 atm.

Note that if the pressure drops suddenly for some reason, the boiling temperature of the water will also drop. This means that if the water is then at a temperature above its boiling temperature it will boil *very* vigorously, changing from a liquid to a gas extremely rapidly. The cooling properties of steam are not as good as those of water, so that the fuel will not be so well cooled. This has very important safety implications for nuclear reactors, as we shall see later.

Question 3.1 A nuclear reactor is run with light-water coolant at 310 °C, and operates at a pressure of 150 atm. The reactor starts to lose pressure at the rate of 5 atm per minute. Assuming that the coolant temperature remains constant, use Figure 3.4 to find out how long it will take before the water starts to boil. In order to do this, you should find what pressure corresponds to a boiling temperature of 310 °C, and then work out how long it takes for the pressure to fall to this value.

Pressurization is also valuable when gaseous coolants are used. At atmospheric pressure, gases have much lower densities than liquids, so they do not transfer heat away from a hot surface very well. However, if we can increase the density of the gas by pressurization, the heat transfer properties improve. Pressurization of the coolant is, in fact, an important feature of all thermal reactors.

Emergency core cooling

In a nuclear reactor, the energy released during nuclear fission is removed by the coolant in the form of heat.

▷   Is fission the only process in which heat is produced in nuclear fuel?

▶   No; heat is also produced in the fuel when fission products undergo radioactive decay (Section 2.2.1).

This heat is called **fission product heating**. Why is this important? As we saw in Section 2.5.3, we can stop the chain reaction, and hence fission of the fuel, using a neutron absorber, but we *cannot* stop the radioactive atoms decaying. So, once the fuel contains radioactive fission products, heat will still be produced in it *even if the reactor is shut down*.

The *amount* of heat will depend on how long the fuel has been in the reactor. The fission products accumulate while the fuel is in the reactor so that, if the fuel is nearly new, the fission product heating will be very small; on the other hand, when the fuel is near the end of its life (that is, it is about to be replaced), the fission product heating will be at its highest.

Of course, since the fission products are radioactive, if the reactor is shut down, or if the fuel is removed from the reactor, the amount of heat produced by them will drop as they decay.

Figure 3.5 shows how the heat produced by the fission products varies with time after the **shutdown** of a reactor that produced 3 000 MW of heat (corresponding to about 1 000 MW of electricity).

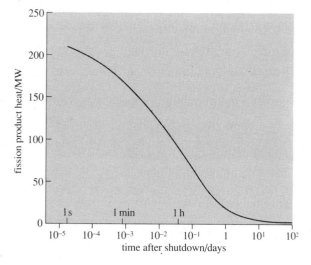

Figure 3.5 Fission product heating after shutdown of a reactor producing 3 000 MW of heat. The fuel is assumed to be halfway through its life. Note that the time axis has a logarithmic scale; because there is no zero on this scale, the fission product heat at shutdown (that is, the moment when the chain reaction ceased) can be approximated to that at 1 second.

▷   From Figure 3.5, how much fission product heat is being produced at the instant of shutdown?

▶   About 210 MW.

▷   How much heat is being produced 1 day after shutdown?

▶   About 20 MW; as noted above, the heat production decreases with time.

If this heat is not removed, the fuel may melt. It follows that, once consumption of the fuel has begun, it must be cooled at *all* times, irrespective of whether or not the reactor is producing energy from fission. Hence, if the main cooling system were to fail, then alternative methods of circulating coolant would have to be provided. Such systems are called *emergency core cooling systems*.

3.1.3 Moderator

A *moderator* is a substance that slows down the fast neutrons (1–2 MeV) produced during fission to thermal energies of about 0.025 eV, at which energy the neutrons have the greatest chance of inducing fission in ^{235}U. As noted in Section 2.5.2, the moderator *surrounds* the fuel rods and consists of nuclei of low mass number.

▷   Why must the moderator contain an isotope with a low mass number?

▶   In Section 2.3.2 you saw that it has to have a low mass number so that a neutron can lose a high proportion of its kinetic energy in a single scattering collision.

The three most important isotopes used as moderators were stated to be hydrogen ($^1_1$H), deuterium ($^2_1$D) and carbon (principally as the isotope $^{12}_6$C), in the form of light water (H$_2$O), heavy water (D$_2$O) and graphite, respectively.

▷   Boron has a lower mass number than carbon, yet it is not in the list of moderators. Why do you think this is?

▶   Boron is not used as a moderator because the boron isotope $^{10}_5$B absorbs neutrons, thus removing them from the chain reaction, and engages in the following charged particle emission reaction:

$$^{10}_5\text{B} + ^1_0\text{n} \longrightarrow ^7_3\text{Li} + ^4_2\text{He} \tag{2.11}$$

Figure 3.6 shows the relative positions of the moderator and other components of a typical thermal reactor. The thickness of moderator needed to separate the fuel elements varies. If the moderator is H$_2$O, only a few millimetres are required between the fuel elements, whereas for D$_2$O and graphite 100 mm or more is necessary. For this reason, the size of the core depends on the moderator being used. Hence the cores of reactors using light water as a moderator are much smaller than those using either D$_2$O or graphite.

3.1.4 Pressure vessels and shielding

Pressure vessels

When coolants were being discussed, you will remember that both liquid and gaseous coolants are pressurized. To enable the coolant to be pressurized, the reactor core is surrounded by a **pressure vessel** as in Figure 3.6.

Shielding

We said earlier that the radiations emitted by radioisotopes are dangerous. The fission products constitute the largest source of radioactivity in a nuclear power station, and it is important that the people working round the reactor are properly shielded. Although the pressure vessel we have just discussed may provide *some* protection, for most reactor types it is not enough, and an additional shield is provided which is separate from this. The material that surrounds the reactor core to provide this protection is called the **biological shield**, or often simply the **shield** (see Figure 3.6).

▷   α-particles, β-particles, γ-rays and neutrons are emitted either during fission or following radioactive decay of the fission products. Which are the two most penetrating of these emissions?

▶   Neutrons and γ-rays are the most penetrating (Figure 2.4).

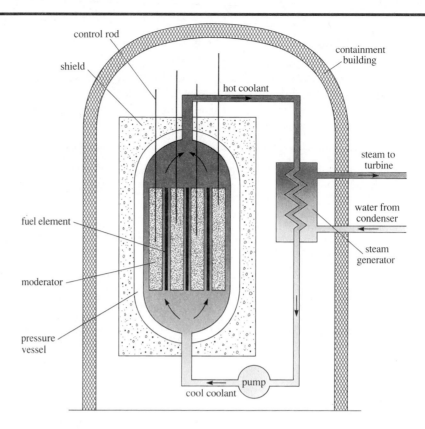

Figure 3.6 Schematic section through a complete reactor system with an indirect steam cycle. Note that the moderator and coolant are different materials in the reactor shown, and so are separate.

When the reactor is running—that is, critical—neutrons will be produced from fission of ^{235}U. However, when the reactor is shut down, and the chain reaction ceases, no neutrons will be produced. On the other hand, γ-rays will be produced both during the fission process and from radioactive decay of the fission products. γ-rays will therefore be produced all the time, although the *intensity* will fall after the reactor has been shut down. The biological shield has to be thick enough to protect the people operating the reactor from these two penetrating radiations at all times. The mixture of elements in concrete provides a very good shield for both neutrons and γ-rays. The thickness required depends on the reactor concerned, but is typically two or three metres.

3.1.5 Reactor containment

▷ What is the function of the cladding round nuclear fuel pins?

▶ The main function of the fuel cladding is to contain the radioactive fission products.

▷ If, for some reason, fission products *do* escape from the fuel pins, do they enter the environment?

▶ Fission products escaping from the fuel pins will enter the coolant. The coolant flows in a *closed loop* (Figure 3.6), so that usually the fission products cannot escape into the environment.

However, what happens if fission products do enter the coolant *and* there is a leak in the coolant circuit? In this circumstance, radioactivity will then escape into the environment around the reactor. If this happens, then the building surrounding the reactor provides the barrier that prevents the escape of fission products into the out-side world. This is called the **reactor containment building** (see Figure 3.6).

3.1.6 Reactor control and instrumentation

In Section 2.5.3 we saw that, in order to control the chain reaction, a material that absorbs neutrons can be used. If the reactor were critical, then the addition of a neutron absorber would make the reactor become *sub*critical. Conversely, if the reactor is shut down—that is, made subcritical—it may be made critical, or *super*critical, by reducing the amount of neutron absorber present.

In practice, changing the reactor between these different states (critical, supercritical or subcritical) is achieved by using **control rods**, made from a neutron absorber, which are able to move in and out of the core, so varying the amount of neutron absorption. Boron is one of the materials used for control rods because of the reaction shown in Equation 2.11, but cadmium is also widely used.

In a reactor there may be fifty or more control rods, spread throughout the core, and these are used to regulate the chain reaction on a minute by minute basis; they are moved in and out while the reactor is running. For the sake of clarity, just four control rods are shown in Figure 3.6. Besides the control rods, there are also **shutdown rods**, absorbers which are either *out* of the core (when the reactor is running) or *in* the core (when it is shut down). These shutdown rods normally work independently of the control rods. They can be used to bring the chain reaction to an abrupt stop in the event of an emergency.

▷ What happens to the rate at which energy is produced from fission when a reactor is made (i) supercritical, and (ii) subcritical?

▸ When a reactor is made supercritical, the rate of energy production from fission increases, and when it is made subcritical the rate of energy production decreases.

When we talk about *controlling* a reactor—apart from shutting it down—we mean determining how much power it produces. Figure 3.7 shows how the control rods are used to do this.

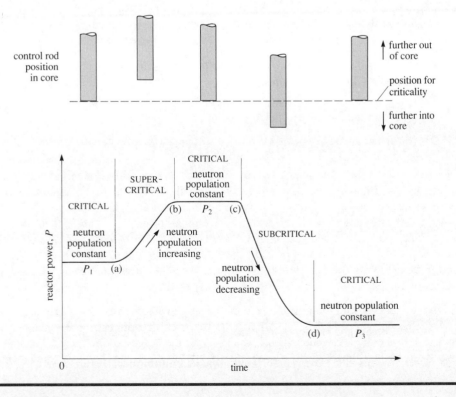

Figure 3.7 Controlling the power produced by a nuclear reactor. To begin with, the reactor, is critical at a power level P_1. When, at (a), there is a need to increase the power, the control rods are withdrawn slightly, the neutron population begins to increase, the reactor becomes supercritical and the power rises. At (b), the control rods are lowered to stabilize the neutron population and to restore the reactor to a critical state at a higher power level P_2. At (c), control rods are inserted further into the core, the neutron population decreases, the reactor becomes subcritical and the power level begins to drop. Finally, at (d), the control rods are withdrawn somewhat to stabilize the neutron population and to restore the reactor to a critical state of lower power level P_3.

The rate at which the fission process is occurring in the reactor core can be monitored by measuring the number of neutrons present in different parts of the reactor. The temperature and pressure of the coolant will be two of the other quantities which are also measured. During normal operation the results of these measurements will be used, for example, to alter the control rod positions and coolant flow rates; however, if the instruments indicate that there may be a fault, the shutdown rods will be inserted, and the chain reaction stopped.

We have talked in terms of *rods* of absorber being used to control, or shut down, a reactor, because this is usually the most convenient way to do so. However, *any* suitable form of absorber can be used, and some reactors use a neutron-absorbing liquid or gas as an extra shutdown system. We shall see later that absorbers in liquid form are used as part of the safety system in some reactors.

Summary of Section 3.1

1   A reactor core consists of rods or pins of nuclear fuel (that is, containing a fissile isotope), encapsulated in cladding to contain the fission products. A fuel element consists of a number of fuel pins, and a complete reactor core contains several hundred fuel elements.

2   The heat produced by fission is taken away by a coolant. The coolant is in contact with the cladding, and flows through the fuel elements. The higher the temperature of the steam entering the turbine the greater is the efficiency of electricity generation, so the coolant that produces the steam should have as high a temperature as possible. The coolant in a water-cooled reactor is therefore pressurized to increase its boiling temperature.

3   In thermal reactors, the fuel elements are surrounded by a moderator, the function of which is to slow down neutrons to low energies, where the probability of fission in the fissile isotope is highest.

4   The power produced in the reactor is regulated using control rods, and the chain reaction is stopped using shutdown rods.

5   Because of the heat produced by fission-product decay, the cooling of a reactor core must continue even after the chain reaction has been shut down.

6   The complete core is surrounded by a concrete biological shield, to minimize the radiation exposure to the operating staff.

3.2 Nuclear power reactor systems

You will now see how the scientific principles that we have discussed come together in the design of real nuclear power stations. Currently (1992), there are eight different types of nuclear power reactor operating throughout the world. However, we are not going to attempt to describe all of them; we are going to discuss three of them in some detail, and give a quick, thumb-nail, sketch of a fourth. A convenient summary of the important characteristics of these four reactors is provided in Table 3.2 at the end of this chapter. You may find it useful to refer to this as an *aide memoire*, or as a means of comparing the reactors with each other, as you read.

Of the reactors we are going to talk about, one, the advanced gas-cooled reactor (AGR), has only been built in the UK; in the course of that discussion we shall compare it with the other common British reactor type, the Magnox. The second has only been built in the former USSR, namely the RBMK or 'Chernobyl' reactor (the letters are a transcription of the Russian initials). The third, the pressurized-water reactor (PWR), is the reactor type most widely used world wide. Finally, our brief description will be of the fast-breeder reactor (FBR), which is important because its use would greatly increase the energy potential of the world's uranium resources.

3.2.1 The advanced gas-cooled reactor (AGR)

The background to the AGRs

Advanced gas-cooled reactors supply over half of the nuclear electricity produced in the UK. The AGR was built as the successor to the Magnox reactor, the first generation of nuclear power reactors in the UK. The Magnox reactors were built in the 1960s, and they are now gradually being phased out. We shall discuss some aspects of Magnox reactors briefly, so that you can see what stimulated the development of the AGRs.

When the Magnox reactors were built, natural uranium (with 0.72% ^{235}U) was the only fuel that was readily available, because enriched uranium could be obtained only from the USA. (Techniques for achieving ^{235}U enrichment are discussed in Section 5.2.1.) As you already know, if natural uranium is used as the fuel, it is difficult to establish a chain reaction unless the absorption of neutrons in other reactor components is kept as low as possible. This limited the choice of the cladding material that could be used. The magnesium–aluminium alloy chosen (called magnox—hence the name of the reactor type) has very low neutron absorption, but has the disadvantages that its melting temperature is low (645 °C) and that it burns in air above 525 °C. To avoid approaching either of these temperature limits, the temperature of the coolant has to be kept below about 400 °C. As a result, the temperature of the steam produced is also low, and so the efficiency of the Magnox reactors is only about 30%. One objective in building the AGRs was to raise this efficiency.

Finally, the use of uranium *metal* as the fuel for the Magnox reactors imposed a limit of 665 °C on the temperature of the fuel, because it swells at higher temperatures due to a change in its metallic structure. Such swelling could rupture the cladding. To overcome these problems, the fuel in an AGR is uranium dioxide (UO_2). This can be made into a ceramic—that is, fused together at high temperatures like pottery—and its melting temperature is over 2 000 °C. The cladding used is stainless steel, which overcomes the low melting and combustion temperatures of magnox alloy, but introduces a new disadvantage: stainless steel absorbs neutrons more strongly than magnox alloy.

▷ What can be done to the fuel to overcome the problem of neutrons being absorbed in the cladding?

▶ The fuel can be enriched in ^{235}U in order to increase the likelihood that the neutrons will be absorbed by fissile nuclei.

An AGR fuel element therefore consists of 36 fuel pins, each of which contains ceramic pellets of UO_2 made from enriched uranium, containing about 2% ^{235}U, and clad in a thin stainless steel sheath (Figure 3.8).

Now let us turn to the coolant. As the reactor's name suggests, it is a gas. It must absorb as few neutrons as possible, and not react chemically with the other reactor components. There are only two gases that have these properties, carbon dioxide (CO_2) and helium (He), and it is carbon dioxide which is used in the AGR.

The AGR is a thermal reactor; that is, it has a moderator, namely carbon in the form of graphite.

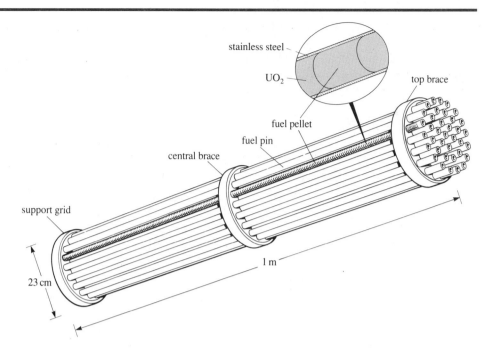

Figure 3.8 A fuel element for an AGR.

▷ Why is carbon a good moderator?

▶ Carbon is a good moderator because its atoms have a low mass number (recall Activity 2.1), and it does not absorb neutrons (cf. Equation 2.11).

The control rods in an AGR are made from a steel alloy containing boron.

The core of an AGR consists of blocks of graphite through which cylindrical **cooling channels** (or **fuel channels**) are cut. The fuel elements are placed in the channels, and the CO_2 coolant flows through the channels (Figure 3.9). There are about 300 fuel channels in an AGR. The complete core is cylindrical, with a diameter of 9.5 m and a height of 8.5 m; the heat output is around 1 500 MW.

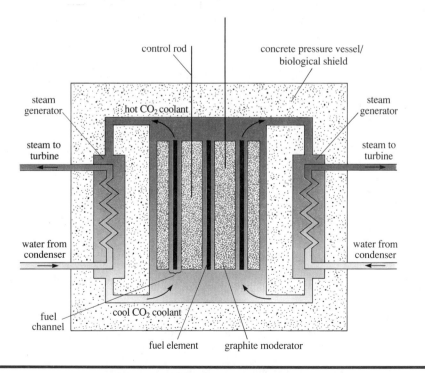

Figure 3.9 Section through an AGR.

The **power density** is one parameter of interest in power reactors, because it is related to the choice of coolant. It is the total heat produced per unit volume of core, and is normally expressed in $kW\,l^{-1}$ or, in what is numerically the same, $MW\,m^{-3}$ (1 cubic metre $\equiv$ 1000 litres).

Let us now work out the power density of an AGR. First we calculate the core volume. As the core is cylindrical,

$$\text{core volume} = \pi \times \text{height} \times (\text{radius})^2 = \pi \times 8.5\,\text{m} \times \left(\frac{9.5}{2}\text{m}\right)^2 = 602\,\text{m}^3$$

As the heat output of an AGR is 1 500 MW, the power density

$$= \frac{1\,500\,\text{MW}}{602\,\text{m}^3} = 2.5\,\text{MW}\,\text{m}^{-3} \text{ or } 2.5\,\text{kW}\,\text{l}^{-1}$$

Let us put this power density into perspective by comparing it to the power density of a normal domestic kettle. These typically have a capacity of 1.5 l, and use 2.2 kW of power. The corresponding power density is $(2.2/1.5)\,kW\,l^{-1}$, that is, $1.5\,kW\,l^{-1}$. So, the power density in an AGR is only about twice that of a domestic kettle. If the kettle were half filled, the power density would be the same as that in an AGR.

▷   What is the function of the pressure vessel in an AGR?

▶   The pressure vessel pressurizes the coolant, in order to increase its efficiency in removing heat (Section 3.1.4).

The pressure vessel in an AGR is made of prestressed concrete, with just a thin steel liner to prevent the coolant gas from escaping. The concrete is 4–6 m thick, which is enough to provide all the biological shielding needed against neutrons and γ-rays.

After being heated in the core, the CO_2 coolant is pumped to one of six steam generators, which are situated in the walls of the prestressed concrete pressure vessel (Figure 3.9). Steam is produced in these steam generators, and the coolant, now at a lower temperature, is pumped back into the core. The steam cycle is therefore an *indirect* one. The steam produced, which has a temperature of about 540 °C, then goes to a steam turbine, and the rest of the electricity production process is exactly the same as in Figure 3.1. Because the steam temperature is so much higher than for the Magnox reactors, the efficiency of an AGR is similar to that for a fossil-fuelled power station, being about 40%.

The location of British Magnox and AGR power stations is shown in Figure 9.5 (p. 191).

Now review the content of Section 3.2.1 by checking that its detail corresponds to that given in items 1–11 of Table 3.2 under the heading AGR.

3.2.2 The pressurized water reactor (PWR)

In 1992, over half of the world's nuclear reactors were of the pressurized water type. They are thermal reactors which use normal, light water (H_2O) as both coolant *and* moderator. However, the hydrogen in light water absorbs such a high proportion of the neutrons produced in fission that a chain reaction cannot be established using natural uranium of any form as fuel. The PWR therefore uses fuel enriched to about 3% in ^{235}U.

As in the AGR, the fuel is uranium dioxide in ceramic form, and a 3.5 m-long stack of 12 mm diameter and 15 mm-high cylindrical pellets, clad in a zirconium alloy forms the fuel pin. Zirconium has the advantages that it resists corrosion by water and does not absorb neutrons to any significant extent. The complete core, which is typically 3.4 m in diameter and 3.6 m high, consists of around 200 fuel elements, each one having between 200 and 300 fuel pins arranged in a square array. The total heat output is around 3 000 MW. The control and shutdown rods, containing silver, indium and cadmium, move up and down within the fuel elements, some of the fuel pins being absent to make this possible.

Because of the need to raise the boiling temperature of the coolant, the whole core is encased in a very thick (200 mm or more) steel pressure vessel. The coolant is at a pressure of about 150 atm, and operates at a temperature of about 320 °C. However, you will remember that the efficiency of electricity production at this steam temperature is only about 30%.

Question 3.2 A PWR produces 1 000 MW of electricity with an efficiency of 31%. Use Equation 3.1 to calculate how many megawatts of heat are produced in the core. The core is 3.4 m diameter and 3.6 m high. Test your understanding of the calculation performed in Section 3.2.1 by working out the power density. How does it compare with that of an AGR?

The main reason for the difference between the AGR and the PWR discussed in Question 3.2 is that light water is a more efficient moderator than graphite, so a smaller mass (or volume) is required; we drew attention to this in Section 3.1.3. Because there is less moderator to heat up, this means that the temperature of a PWR would rise much more rapidly than that of an AGR if the cooling system were to fail.

Unlike in the AGR, the PWR has no cooling channels, and the fuel elements are packed close together. The coolant passes up through the core between the fuel pins, and is then pumped to a steam generator. Having given up as much of its heat as possible, the coolant is then pumped back into the core; thus, the reactor has an *indirect* steam cycle, and the coolant flows in a closed loop without going to the steam turbine. Each reactor has three or four cooling loops, just one of which is shown in Figure 3.10.

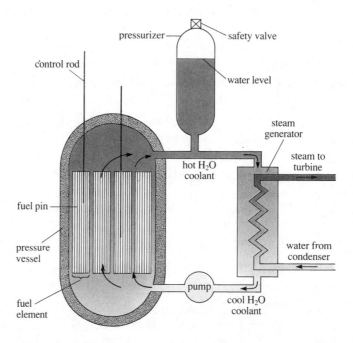

Figure 3.10 Section through a PWR, showing the reactor cooling circuits. There is no separate moderator because light water performs the functions of both coolant and moderator.

In order to maintain a constant coolant pressure, there is a **pressurizer**. Not only does this regulate the pressure, but it also contains a **pressure relief valve**, or valves, which releases the pressure if it rises too much. The pressurizer is an extremely important unit, and its proper operation is *essential* for the safety of the whole nuclear plant. We shall see what can happen if it goes wrong when we discuss the Three Mile Island accident, in Chapter 6.

The reactor and its cooling circuits are all housed in a containment building, in the way shown in Figure 3.6. The PWR containment building (not shown in Figure 3.10) is a fairly substantial one, with concrete walls about 1 m thick. Its purpose is to prevent any radioisotopes that escape from the reactor from entering the environment. But why does it have to have such thick walls?

Remember that the water in the primary cooling circuit has a temperature of over 300 °C. This is only possible because it is under a pressure of 150 atm. If the cooling circuit is ruptured, the pressure drops steeply. This will cause the water to boil and turn to steam almost instantaneously. This transition from liquid to gas is associated with a huge increase in volume, since a litre of water turns into about 200 litres of steam at atmospheric pressure.

The result would be a rise in pressure, which might rupture the containment building. Consequently, the containment building for a PWR has thick walls, on the inside of which there are cooling units for the *building* (cold-water sprays, fans or refrigerators), which condense any steam released, and keep any increase in pressure inside the building to a minimum.

We did not mention the containment building when we discussed the AGR because apart from being more leak-tight, it is similar to a normal industrial building. This is possible because the reactor and its associated plant are much larger than a PWR. The building around an AGR is therefore also larger, and any escaping gas would not change the pressure inside the building significantly.

Now review the content of Section 3.2.2 by checking its consistency with items 1–11 of Table 3.2 under the heading PWR.

3.2.3 RBMK: a light-water cooled, graphite-moderated reactor system

RBMK is the acronym for the type of reactor that failed at Chernobyl in 1986. Although there is an 845 MW (electrical) power station of similar design at Hanford, Washington, USA, all other examples of this type of reactor have been built in the former USSR. As we shall see, it is the *combination* of features which makes the design unique, not the features themselves.

RBMK is a thermal reactor and, like the AGR and PWR, the fuel is enriched uranium (about 2% ^{235}U) in the form of ceramic UO_2. The coolant is light water, pressurized to 70 atmospheres to increase its boiling temperature, and the moderator is graphite. The core diameter is just over 12 m, and it is 7 m high; the total heat produced is 3 200 MW, and the electrical output is 1 000 MW. The power density is 3.8 kW l^{-1}, which is comparable with that of an AGR.

▷ What is the efficiency of an RBMK reactor?

▶ The efficiency is given by Equation 3.1:

$$\frac{\text{electrical power out}}{\text{heat produced}} \times 100\% = \frac{1\,000\,\text{MW}}{3\,200\,\text{MW}} \times 100\% = 31\%$$

Thus, the efficiency of an RBMK reactor system is very similar to that of a PWR. This is because both use light-water coolant.

So far, therefore, the RBMK reactor has features of both the PWR and AGR reactors. However, the way in which the coolant is pressurized is totally different from either of them. Each fuel element consists of 18 fuel pins clad in a zirconium alloy, as in the PWR. The fuel elements are set in fuel channels consisting of a zirconium alloy tube, through which the pressurized coolant flows. There are 1 700 of these **pressure tubes** running vertically through the graphite moderator, and, together they comprise the pressure vessel (Figure 3.11).

The graphite moderator is therefore separated from the light-water coolant, and the moderator temperature is about 700 °C under normal operation. The control rods, made from boron carbide, run between the pressure tubes containing the fuel elements.

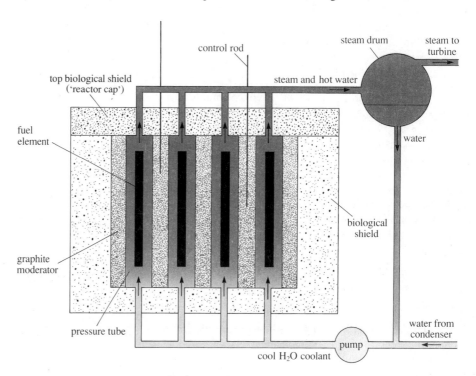

Figure 3.11 Section through the core and cooling circuit of an RBMK reactor.

Apart from the mode of pressurization, another difference between the RBMK reactor and the PWR is that, although the light-water coolant is pressurized to raise its temperature, it is allowed to boil in the core. The steam is then fed through steam drums (Figure 3.11), which separate out any water droplets, and then on to the turbines; the reactor has a *direct* steam cycle. (In contrast, the reactor at Hanford mentioned earlier uses an *indirect* steam cycle.)

The steam drums and cooling pumps are all housed in separate reinforced concrete vaults. The reactor itself is in its own vault, but the principal barrier between the reactor and the external environment is a 3 m thick, 2 000 t reinforced concrete biological shield ('reactor cap'), which rests on the side walls of the main biological shield. At the bottom of the vault housing the reactor is a very large pool of water, which serves the same purpose as the cooling units in a PWR building.

▷ What is the function of the cooling units in a PWR building?

▶ The cooling units are used to condense the steam that is produced if the cooling loop leaks. This lowers the steam pressure in the building, and helps to prevent damage to it.

The building surrounding the whole RBMK reactor and its associated equipment is similar to that round an AGR, and is not built to withstand high pressures; that is, there is no containment building. As with *all* reactor systems, the turbines are not housed in the same building as the reactor, but separately.

Now review the content of Section 3.2.3 by checking its consistency with items 1–11 of Table 3.2 under the heading RBMK.

Summary of Section 3.2

1   AGRs have only been built in the UK, and were designed to overcome the limitations of the earlier Magnox reactors. AGRs use an enriched (2%) ceramic UO_2 fuel, clad in stainless steel. The moderator is graphite, and the coolant is CO_2. AGR reactors incorporate a prestressed concrete pressure vessel, which also serves as a biological shield. They have an efficiency of about 40%, which is comparable to that of a fossil-fuelled power station.

2   The fuel in PWRs is UO_2 containing enriched (3%) uranium, clad in a zirconium alloy. Both the moderator and the coolant are light water. The core is comparatively small, and the power density is over $80 \, kW \, l^{-1}$, which is much higher than for an AGR. They have a thick steel pressure vessel, and the reactor and steam generators are housed in a special containment building. PWRs are the dominant reactors world wide.

3   The RBMK reactor is unique to the former USSR, although most of its features are to be found in other reactors. It uses enriched (2%) UO_2 fuel, clad in a zirconium alloy. The coolant is light water, which boils in the core, and the steam produced drives the turbines directly. The coolant is pressurized using pressure tubes, which form the cooling channels. The moderator is graphite.

3.3 Fuel requirements for nuclear reactors

So far, we have discussed nuclear reactors without saying anything about the availability of the fuel they use. The only naturally occurring fissile isotope is ^{235}U, which forms 0.72% of natural uranium. So how much uranium is there in the world?

This question is difficult to answer accurately. It is normal to describe uranium resources in terms of those which are 'reasonably assured'—that is, have been well explored—and those for which a good estimate can be made, the so-called 'estimated additional resources'. In the case of uranium, the world supplies of reasonably assured uranium resources which could be economically extracted were estimated to be around 2 300 000 t in 1988. In addition, the estimated additional resources comprised a further 1 300 000 t.

Now the amount of natural uranium required over the 30-year lifetime of a reactor producing 1 000 MW of electricity varies between the reactor types, but is around 5 000 t. (Notice how small this is compared to the amount of coal required by a fossil-fuelled power station of the same size, namely about 3 000 000 t *per year*, or 90 000 000 t over 30 years.)

▷   If *all* the reasonably assured uranium resources were used, how many reactors would this provide fuel for over their complete working lifetime?

▶   If we assume a figure of 5 000 t of natural uranium as a lifetime requirement per reactor, then the reasonably assured resources would fuel (2 300 000/5 000) or 460 reactors.

There are already 439 reactors in the world, of which 285 are shown in Table 3.2. So by the time they have all reached the end of their lives (around the year 2020), a substantial proportion of the world's reasonably assured supplies of uranium will have been used up. The nature of exploration and mining is such that this would not happen in practice. As demand increases, so does the effort put into further exploration, and it is certain that uranium resources will be extended in this way. Nevertheless the point about limited resources is still true; using thermal reactors of the type currently in use, the contribution that nuclear power could make to the world's energy needs is limited. Indeed, of all the non-renewable fossil-fuel energy sources (coal, oil and gas), only coal can be considered as a long-term resource, as Table 3.1 shows. Even with coal, the estimated lifetime, of 250 years at current usage rates, is short on historical time-scales.

Table 3.1 Estimated availability of non-renewable fuels and their estimated lifetimes at 1987 use rates.

Fuel*	Proved recoverable resources (1987)	World annual usage (1987)	Fuel lifetime at 1987 use rates /years
coal	799×10^9 t	3.2×10^9 t	250
natural gas	106×10^{12} m^3	1.7×10^{12} m^3	62
crude oil	123×10^9 t	2.8×10^9 t	44
uranium	2.3×10^6 t	37×10^3 t	62[†]

* Source: 1989 Survey of Energy Resources, World Energy Conference, published in 1989 *Energy Statistics Yearbook*.
[†] In thermal reactors using ^{235}U.

Data like those in Table 3.1 are always questionable because they are estimates assuming a usage that is fixed at the current level. This is unlikely to be the case. Nevertheless, the *broad* conclusions that can be drawn from the data in this table are unlikely to be wrong. One conclusion, which springs from the relatively short lifetimes estimated for *all* fuels, is that the case for energy conservation, and for the exploitation of the renewable resources (such as solar, wind, tidal and wave power) in the long term, is unquestionable.

As regards nuclear power, if supplies of the fissile isotope ^{235}U that is used in thermal reactors are so limited, how can nuclear power contribute substantially to future world energy demands? The answer is that ^{238}U can be made to absorb neutrons in a nuclear reactor and decay to the fissile isotope ^{239}Pu. If all ^{238}U were converted in this way, the amount of fissile material would be increased some 140-fold, as Activity 2.2 showed. In practice, the conversion is not perfectly efficient, and a factor of 60 would be more realistic. Even so, this is an enormous increase, and transforms nuclear power into an energy resource of greater importance than coal, particularly since it may then be economic to mine grades of uranium ore which are more costly than the grades currently exploited for thermal reactor fuel. In Activity 2.2, you also showed that natural thorium, ^{232}Th, can likewise be converted to the fissile isotope ^{233}U in a reactor. The world's thorium supplies are even greater than those of uranium, so that breeding from thorium would substantially increase the nuclear energy potential.

This, then, is the background against which we have to consider breeding fissile isotopes for nuclear fuel. Although the use of thermal reactor breeder systems using thorium has been proposed, the conversion of ^{238}U to plutonium is the one that has been most widely explored, and we shall restrict our discussion to this.

Summary of Section 3.3

1   Assessments of world uranium resources suggest that, if it confines itself to the fission of ^{235}U in thermal reactors, then at present rates of consumption, the nuclear industry will only be able to supply energy for a few decades.

2   This span of time could be increased some 60-fold by using breeder reactors in which ^{239}Pu is made from ^{238}U, and by an even larger factor if ^{239}Pu breeding were supplemented by ^{233}U breeding from ^{232}Th.

3.4 Fast-breeder reactor systems

Fast reactors differ from thermal reactors in a number of ways. For one thing, they have no moderator.

▷   How will this affect the energy of neutrons emitted during fission?

▶   The function of a moderator is to slow down the neutrons emitted during fission. This maximizes the probability of a fission reaction occurring in fissile isotopes such as ^{235}U. If there is no moderator, then the neutrons will not be slowed down. Their average kinetic energy will therefore be much higher than in a thermal reactor.

However, as you saw in Section 2.4, the chance that neutrons will induce fission in ^{235}U is far lower at high neutron energies than at low ones (below 1 eV).

▷   How is it possible to achieve criticality when the probability that fission will occur is so low?

▶   By increasing the number of fissile atoms present in the fuel.

Thus, in a fast reactor the percentage of fissile isotope in the fuel is increased to 20 to 30% compared with the 0.72 to 3% figure generally used in thermal reactors. This greatly increases the chance that a neutron will encounter a fissile nucleus in the fuel.

As a consequence of having no moderator, the core of a fast reactor is very small, being only about 2.5 m in diameter and 1.5 m high for a reactor producing 1 000 MW of electricity. This means that the power density is far higher than in thermal reactors; assuming an efficiency for electricity production of 40%, the power density is about 340 MW m^{-3} (340 kW l^{-1}). Putting about 200 one-kilowatt electric fires into a pint beer glass would achieve the same result, but only someone who had drunk a lot of beer would attempt it!

To remove such large amounts of heat from such a small volume, the coolant clearly has to be very special indeed. The only *proven* coolants are liquid sodium or some of its alloys, which have a very high thermal conductivity, although high-pressure helium has also been proposed.

One *advantage* of liquid sodium as a coolant is that its boiling temperature at atmospheric pressure is high (881 °C). Because the reactor operating temperature is much lower than this (normally around 600 °C), the coolant does not have to be pressurized to keep it from boiling. However, a risk associated with the use of sodium is that it reacts very vigorously with water to produce hydrogen gas (Equation 3.2), a reaction that can proceed with explosive force:

$$2Na(s) + 2H_2O(l) \longrightarrow 2NaOH(s) + H_2(g) \qquad (3.2)$$

This means that water *must* be kept out of the core. To help achieve this, *two* cooling circuits are used, the liquid sodium cooling the core giving its heat up to liquid sodium in a second cooling circuit, which then goes to a steam generator (Figure 3.12).

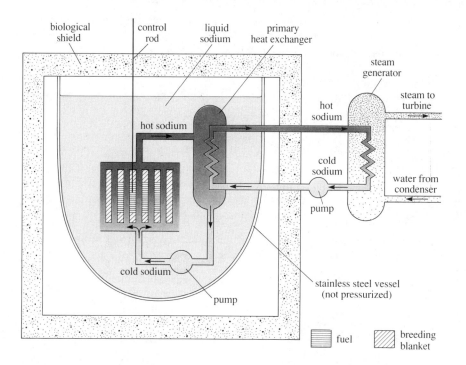

Figure 3.12 Section through a sodium-cooled, fast-breeder reactor system.

Apart from the absence of a moderator, the reactor core has features similar to the thermal reactors described so far, although the fissile isotope used is ^{239}Pu, not ^{235}U. The fuel (20–30% ^{239}PuO$_2$ intimately mixed with ^{238}UO$_2$) is in the form of small pins, 5 mm in diameter, clad in stainless steel, a fuel element having 625 pins. The pins are separated by about 0.5 mm, to allow the coolant to flow. The whole core contains about 200 fuel elements, packed together as in a PWR, and liquid sodium coolant is pumped up through the fuel elements.

As with the AGR, the fast-breeder reactor uses control rods made from a boron steel.

One big difference from thermal reactors is that the core is surrounded by special stainless steel-clad pins containing ^{238}U. This is called the **breeding blanket**, since it is where most of the production ('breeding') of the new fissile material occurs. A fast reactor in which such breeding occurs is known as a **fast-breeder reactor (FBR)**. Activity 2.2 dealt with the breeding process; for ^{239}Pu the three-step sequence is:

$$^{238}_{92}\text{U} + {}^{1}_{0}\text{n} \xrightarrow[\text{capture}]{\text{radiative}} {}^{239}_{92}\text{U} \xrightarrow{\beta\text{-decay}} {}^{239}_{93}\text{Np} \xrightarrow{\beta\text{-decay}} {}^{239}_{94}\text{Pu} \qquad (2.12)$$

Although some of the plutonium formed in the breeding blanket will undergo fission, the majority of the power produced comes from fissions occurring in the plutonium fuel in the reactor core. Indeed, the thermal energy produced per unit volume is so high in a fast reactor core that, if the coolant were to stop flowing, the fuel in the core would melt within seconds. For this reason, as Figure 3.12 shows, the whole core (with the surrounding breeding blanket) is immersed in a pool of liquid sodium; heat can then pass into the pool of sodium even if a pump fails.

Now review the content of Section 3.4 by checking its consistency with items 1–11 of Table 3.2 under the heading FBR.

3.4.1 The rate at which fissile fuel is produced

In Section 2.6 we said that breeding was a process whereby more fissile material was produced than was used as fuel. Let us suppose that we need 1.40 t of fissile material for a breeder reactor core. If this reactor *uses* 0.48 t of fissile material per year, and *produces* 0.60 t, then it will have produced 0.12 t *more* fissile material per year than it used.

▷ How long will it take to produce enough extra fuel to start a new reactor?

▶ If it takes 1.40 t of fissile fuel to start a new reactor, and 0.12 t of excess fuel is produced per year, then it will take (1.40/0.12) years, that is, 11.7 years, to produce enough extra fuel to start a new reactor.

The time it takes to produce enough extra fissile fuel to start a new reactor is called the **doubling time** of the breeder reactor. Note that the original breeder reactor has not only produced this extra fuel, but has also produced as much new fissile material as it used up; that is, the amount of fissile material has been doubled.

▷ Why is the doubling time of breeder reactors important?

▶ Because it determines how quickly a nuclear power programme based on breeder reactors could increase the supply of electricity.

An analogy to animal breeding serves us well here; the shorter the gestation time, the more quickly an animal population can multiply. Even allowing for multiple births, populations of rabbits (gestation time 28 days) increase more rapidly than those of elephants (gestation time over 600 days)!

Question 3.3 A fast-breeder reactor produces 1 000 MW of electricity. As soon as it has produced enough extra fuel, this is used to start a new reactor of the same electrical output. If this process is continued for 40 years, and all the reactors are kept running, what would the total electricity generated be at the end of this period, assuming doubling times of (a) 10, and (b) 20 years?

What you will have found is that the doubling time is the critical factor in determining how quickly an electricity generating programme based on breeder reactors could grow. The doubling times for the prototype fast-breeder reactors under investigation range from 15 to 25 years.

Summary of Section 3.4

1 Fast-breeder reactors have no moderator. A typical fuel consists of 20–30% $^{239}PuO_2$ mixed with $^{238}UO_2$. In the absence of a moderator, the core is very small and the power density very high, so liquid metal coolants (usually sodium) are used.

2 The core is surrounded by a breeding blanket of $^{238}UO_2$, where most of the new ^{239}Pu is formed. The rate at which an electricity generating programme based on breeder reactors can grow is greater the shorter the doubling time of the reactors.

Activity 3.1   *You should spend up to 30 minutes on this activity.*

You have now read a description of the components of nuclear reactors, and seen how these components are put together in four working designs. Now let us see if you can use what you have learnt to make sense of a design that you have not been told about. Figure 3.13 is a schematic and highly simplified diagram of the Canadian CANDU (CANadian Deuterium–Uranium) reactor.

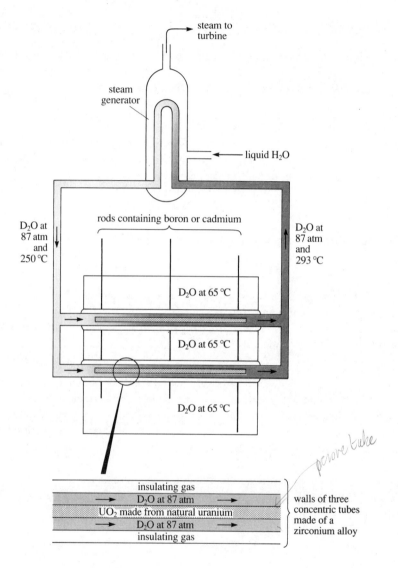

Figure 3.13 Section through the CANDU reactor.

It includes a detail of part of the reactor core. Study the diagram carefully, and then answer the following questions.

(a) What fissile isotope undergoes fission in this reactor? ^{235}U

(b) Why is the fuel in this type of nuclear reactor cheaper than in most other types? not enriched

(c) Is this a thermal or a fast reactor?

(d) Identify the coolant; is it involved in a direct or indirect steam cycle? D_2O

(e) Why is some of the D_2O under a pressure of 87 atm? higher boiling temp

(f) What is the function of the rods containing boron or cadmium? control rods

(g) One of the disadvantages of using light water in a reactor core is the tendency of the isotope ^1H to absorb some neutrons. The chance of a ^2H ($\equiv ^2D$) nucleus doing this are only about one five-hundredth that of a ^{1}H nucleus. How has this affected the nature of the fuel in this reactor? *If fewer n will be adsorbed, more chance of fission → low neutron absorption*

(h) This design resembles the RBMK reactor in having pressure tubes. Locate a pressure tube in the diagram and describe its purpose. *Comprise pressure vessel*

(i) The reactor produces 600 MW of electricity and has an efficiency of 28%. The cylindrical core has a diameter of 6.5 m and a height of 6.0 m. Calculate the power density.

$\frac{600\,MW}{28} \times 100$

$\pi \times 6m \times \left(\frac{6.5}{2}\right)^2 = 199\,m^3$

$= 10.74\ MW\,m^{-3}$

raise temp of coolant to increase efficiency

Activity 3.2

Extract 3.1 is taken from a school science textbook published in 1985. Read it and then answer the following questions:

(a) As both Extract 3.1 and this chapter have acknowledged, estimates of fuel resource lifetimes are subject to many uncertainties, and tend to change with time. Compare the expiry date of uranium resources implied by Table 3.1 with that given in the bar chart in Extract 3.1.

(b) Why does Extract 3.1 grossly underestimate the potential for nuclear energy based on natural uranium? In addition, what further source of nuclear energy also based on nuclear fission does this extract also overlook? *Further*

(c) Starting with the data on uranium given in the bar chart in Extract 3.1, correct it for the factors uncovered in part (b) of this activity and thereby obtain a new uranium resources expiry date.

^{238} Pu
$60 \times 2 \times 20yr$
$= 4400$

Extract 3.1 From *Starting Science*, Book 1, A. Fraser and I. Gilchrist (1985).

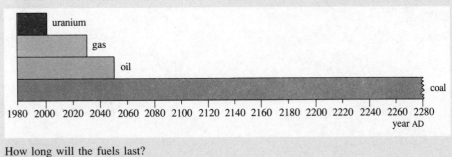

3.3 When the fuels run out

In the 1980s, most of the world's energy will be supplied by fuels—coal, oil, gas and uranium. But these fuels won't last forever. If most experts are right, some fuels will run out in your lifetime.

The bar chart below shows you how long present supplies of fuel are expected to last. Of course, the chart could change. If everyone decided to use fuels more sensibly, they would last longer.

When some of the fuels run out, there could be a serious shortage. That is why scientists and engineers are working hard to find ways of getting useful energy from new energy sources. They are trying to find ways:

- to use the moving energy of the wind and waves to turn generators to make electricity

- to use the sun's heat energy to heat homes, and to make cheap photocells to change its light energy to electricity

- to use the heat energy of 'hot rocks' which lie just below the Earth's surface

- to control the tremendous energy from nuclear fusion reactions which produce temperatures as hot as the sun

- to produce new fuels—like hydrogen from water or alcohol from sugar cane or other plants

By the time the fuels run out, at least some of these problems should have been solved.

How long will the fuels last?

Table 3.2 The principal features of, and typical data for four reactor types.

Feature	Reactor type			
	AGR	PWR	RBMK	FBR
1　fuel type	UO_2	UO_2	UO_2	PuO_2
2　fissile isotope and % in fuel	^{235}U 2%	^{235}U 3%	^{235}U 2%	^{239}Pu 20–30%
3　cladding	stainless steel	Zr alloy	Zr alloy	stainless steel
4　coolant	CO_2	H_2O	H_2O	sodium
5　moderator	C (graphite)	H_2O	C (graphite)	none
6　control rod material	steel alloy containing boron	cadmium alloy	boron carbide	steel alloy containing boron
7　core diameter/m	9.5	3.4	12.2	2.5
core height/m	8.5	3.6	7.0	1.5
8　pressure vessel	concrete	steel	Zr alloy pressure tube	none
9　steam cycle	indirect	indirect	direct	indirect
10　efficiency (%) *	40	31	31	40
11　power density/ $MW\,m^{-3}$	2.5	99	3.8	340
12　number of reactors world wide †	14	239	28	4
The data below are discussed in Chapter 5.				
13　fuel burn-up/ $MW\,d\,t^{-1}$ ‡	18 500	32 000	22 500	80 000
14　total commercial electrical capacity world wide (1990)/MW † §	9 300	212 500	17 800	270

* $\dfrac{\text{electrical energy output}}{\text{heat energy produced}} \times 100\%$

† Source: World Nuclear Industry Handbook 1991, Nuclear Engineering International. The same source gives the total number of nuclear power reactors in the world as 439, and the total commercial nuclear electricity capacity as 341 570 MW

‡The significance of this quantity will be discussed in Section 8.2.1.

§ The world's net installed capacity of electricity generating plant in 1989 was 2.7×10^6 MW.

(Source: *UN Energy Statistics Yearbook* 1989.)

4 The biological effects of radiation

The main waste products from nuclear power generation are radioactive fission products. When these fission products decay, they emit radiation, which can cause biological damage, possibly resulting in illness or death. In this chapter we shall look at how radiation causes biological damage, and at the way in which the limits on the exposure to radiation are decided. We also consider the evidence for the relationship between exposure to radiation and the incidence of leukaemia in children.

4.1 Introduction

Use of the word 'radiation' often invokes a feeling of something mysterious, perhaps even sinister. For example, the famous character from *Marvel Comics*, the Incredible Hulk, was a scientist who became transformed as a result of exposure to radiation! Part of the reason for this feeling is that radiation cannot be detected directly with any of our senses: we cannot hear, see, smell, taste or touch it. Any danger from it is not directly apparent to us. However, its effects are not necessarily malevolent, as countless people who owe their lives to cancer cures through radiation therapy can testify. So what are the properties of radiation, which can be both beneficial and harmful?

There are several different phenomena to which we give the name *radiation*. Electromagnetic radiation is one broad class, which includes visible light, microwaves, X-rays and γ-rays. Neutrons, and the β-particles and α-particles arising from radioactive decay, are examples of other types of radiation. Let us first consider the α- and β-particles.

▷ Why is an atom electrically neutral?

▶ Because it contains equal numbers of positively charged protons and negatively charged electrons.

▷ Are α- and β-particles electrically neutral?

▶ No. α-particles are helium nuclei, with two protons and two neutrons; they therefore carry a charge of +2. β-particles are electrons, and therefore carry a charge of −1.

Between two charged particles there is an electrical force; if the charges are different, the force is one of attraction; if they are the same, the force is one of repulsion. Thus, when an α-particle or a β-particle passes through matter there will be an electrical force between the particle and the electrons in the atoms of the surrounding material. The force can be enough to remove or eject one or more electrons from their orbits, leaving these atoms with a net positive charge. This is a process of **ionization**. The forms of radiation which cause damage to humans do so because they produce ionization, and are called **ionizing radiation**.

The process of ionization just described involves the transfer of energy from the radiation to matter, and results in the removal of one or more electrons. However, in chemical compounds the electrons are the 'glue' that holds the atoms together: covalent bonds, for example, can be thought of as pairs of electrons. Thus, when ionizing radiation removes electrons from living tissue, some bonds may be broken, and others may be formed; that is, chemical changes may occur, which can then produce biological changes.

Hence, it is not the amount of energy transferred to the biological material, which is of paramount importance; rather it is the biological damage caused by the resulting ionization. In fact, the *amount* of energy that is needed to cause damage to biological systems is *minute*. For example, the amount of ionizing radiation needed to kill a fully grown person would transfer only about enough energy to increase the body temperature by 1/400 °C (0.002 5 °C)! On the other hand, the amount of ionization produced *is* proportional to the energy absorbed, and is therefore a *measure* of the biological damage that the radiation causes.

It is important to remember that ionizing radiation is not a twentieth-century phenomenon. Our bodies have to withstand constant exposure to ionizing radiation from a variety of *natural* sources: the sum total of this is called the **background radiation**. Radioactive materials exist in the soils and the rocks around us, and ionizing radiation reaches us from the stars via cosmic rays. However, in the last hundred years or so, new sources of ionizing radiation have arisen from human activities, of which X-radiation is one familiar example. To see how much you already know about exposure to ionizing radiation, you should now do Activity 4.1.

Activity 4.1

The main sources of exposure to the general public from ionizing radiation are:

(i)   fall-out from nuclear weapons;

(ii) medical uses of radiation and radioisotopes;

(iii) cosmic rays;

(iv) rocks and soils;

(v)   inhaled radioactive gases from natural sources (radon);

(vi) discharge of nuclear power wastes;

(vii) natural radioactive substances in food and drink.

(a)   List the above sources in order of the contribution you think that they make to the radiation exposure in the UK. Then compare your answer with Figure 4.1, which shows the figures for the average radiation exposure in the UK.

(b)   Use the pie chart in Figure 4.1 to decide what percentage of the total radiation exposure to the public arises from natural sources. These data are supplied by the (UK) National Radiological Protection Board. Why do you think the radiation exposure *you* receive might be different from this?

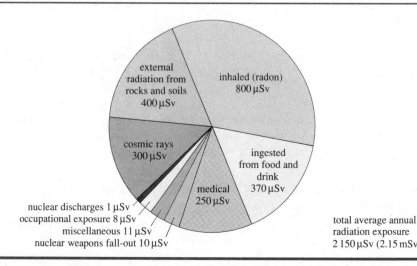

Figure 4.1   The average annual radiation exposure from various sources to the public in the UK. The unit used is discussed in Section 4.3.

4.2 Radiation damage to biological systems

Before we look at the effects of ionizing radiation on humans, we need to consider the effects that such radiation has on biological systems at a cellular level.

Multicellular (and some unicellular) organisms are called **eukaryotes**, in contrast to simpler organisms (such as bacteria), which are called prokaryotes.

▷ What features distinguish the cells of eukaryotes from those of prokaryotes?

▶ There is a distinct **nucleus** in eukaryotic, but not prokaryotic, cells; indeed, prokaryotic cells contain no cell organelles at all.

Figure 4.2 shows the basic structure of a single eukaryotic (in this case, animal) cell. Box 4.1 reminds you of some of the important features of such cells and of how they divide.

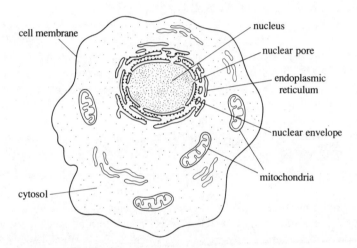

Figure 4.2 The principal features of an animal eukaryotic cell.

Box 4.1 The eukaryotic cell

Eukaryotic cells (whether of animals or plants) consist of a matrix called the **cytosol** (75% of which is water), surrounded by the **cell membrane**. By means of a complex system of delicately balanced channels and chemical pumps, this membrane controls what enters and what leaves the cell. Within the cell the various cellular functions are carried out both within the cytosol and by the **organelles**. Perhaps the most important organelle in the context of radiation damage is the nucleus.

The nucleus contains pairs of **chromosomes** (23 pairs in the case of humans), which are made up of protein and helical strands of **deoxyribonucleic acid (DNA)**. DNA itself is made up of several compounds, notably the four organic bases **adenine, guanine, cytosine** and **thymine**. Each cell's **genes**, which are essentially lengths of its DNA within a chromosome, contain all the information necessary to produce not only a particular cell but the whole organism. This information is encoded in the sequence in which the four bases are arranged along the DNA strand. If this sequence is disrupted in any way, then the 'message' carried by a gene may be changed. Such disruptions are called **mutations**. Mutations can occur at a molecular level, involving small changes in the structure of one of the bases, the deletion or addition of a single base, or the substitution of one base for another. They can also occur on a larger scale, resulting in structural changes to one or more of the chromosomes. These changes are sometimes visible under the light microscope.

▷ Name the two types of cell division.

▶ **Mitosis** and **meiosis**.

4.1

▷ What are the main differences between them?

▶ In the absence of mutation, mitosis results in the production of new **somatic** cells, which are genetically identical to one another and to their parent cell. In contrast, meiosis results in the production of **gametes** (that is, sperm or ova), each containing just one representative of each pair of chromosomes, and so *half* the number of chromosomes present in the parent cell. In addition to this difference in the *amount* of genetic material, the gametes *differ genetically* from their parent cell (and from one another) as a result of genetic recombination (that is, reassortment and crossing-over).

Thus, mitosis is concerned with the growth and repair of individual organisms. Meiosis is an integral part of sexual reproduction; the fusion of two gametes at fertilization produces a genetically unique **zygote**, which then grows as a consequence of mitotic division. ■

Ionizing radiation can cause damage to any of the component parts of a cell, but damage to the nucleus tends to be the most serious. The reason for this is the key role played by DNA.

Consider a group of *somatic* cells, whose DNA is disrupted by ionizing radiation. The cells may be killed or so badly damaged that they may be unable to divide by mitosis. This can have very serious implications if, for example, the cells are part of a tissue that is normally rapidly dividing, such as the skin or the lining of the gastro-intestinal tract. In these tissues, the surface cells protect underlying cells from a harsh environment. They are therefore continuously being detached and replaced by others from their rapidly dividing parent, or stem, cells. If division of these cells is halted or slowed, then the lost surface cells are not replaced. The consequent exposure of the more vulnerable cells underneath to the harsh environment can quickly result in even more serious damage.

Alternatively, although a somatic cell exposed to ionizing radiation may be able to replicate, its DNA may still have undergone a mutation that will not be revealed until later. For example, the mechanism by which cell division is controlled may have been damaged. This can result in the production of cells that do not behave normally. In most cases the body's immune system would recognize these cells as abnormal and destroy them. However, sometimes cells resulting from uncontrolled cell division are not destroyed, and go on to develop into a tumour, or **cancer**. One cancer that is known to be related to high levels of exposure to ionizing radiation is **leukaemia** (the general name given to certain cancers of the bone marrow and lymphatic system).

If a mutation occurs in one of the cells that give rise to *gametes*, the effects may not appear in the irradiated individual, but in any offspring the individual may subsequently have. In these circumstances *all* the somatic cells in the offspring's body (including those that give rise to *their* gametes) will carry the mutation. Although some inherited mutations may be advantageous in certain circumstances, the vast majority are assumed to be disadvantageous.

As living things have always been exposed to ionizing radiation from natural sources, biological systems have developed various mechanisms to cope with or repair damage caused by ionizing radiation. For example, new DNA may be synthesised to replace damaged strands. It is only when the level of radiation damage exceeds the capacity of the repair mechanism to cope that a cell is killed, begins to divide uncontrollably or causes a harmful mutation to be passed on from one generation to the next.

There are two circumstances in which the repair mechanisms can fail to cope. The first is when the energy absorbed by a specified mass of tissue (called the radiation **dose**) following irradiation is very large, and it is transferred very quickly. The other is where the dose from a single irradiation is smaller, but there is a succession of irradiations separated by quite short time intervals. In both cases, the repair systems cannot keep pace with the damage caused; the *rate* at which the energy has been transferred—the **dose rate**—is too high.

The effects of dose rate can be seen if cultures of living cells are irradiated. Along the *y*-axis of Figure 4.3 is plotted the percentage of cells that survive following *two* separate radiation doses. The *x*-axis of the graph shows the time interval between the two exposures. When the interval is long (24 hours), over 80% of the cells survive both exposures. This also tells us that *one* of the exposures by itself, is not sufficient to kill a high percentage of the cells. However, when the period between the exposures is considerably shorter than 24 hours, many cells that have been damaged by the first exposure will not be able to repair themselves sufficiently to survive a second irradiation. For example, when the cell culture is left for only 6 hours between exposures, only 50% of the cells survive.

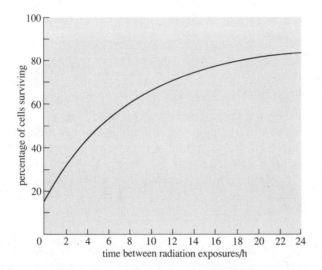

Figure 4.3 The dependence of cell repair on the interval between radiation doses. The cell cultures (collections of cells grown in a special solution, rather than in a living organism) were given two radiation doses of 1 Gy separated by the time interval shown along the *x*-axis and the percentage of cells surviving the second dose was measured. Do not worry yet about the *units* of radiation dose, because these are dealt with in Section 4.3.

Therefore, the survival of cells depends not only on (i) the total radiation dose received but also on (ii) the rate at which that dose is given.

A third factor that influences the biological damage is the *type* of radiation involved. Let us take α-particles first. They are emitted by unstable heavy elements such as plutonium, uranium and polonium. The range (penetrating power) of an α-particle is very short. For example, an α-particle with an energy of 1 MeV will only travel about 10 mm or so in air, and would not penetrate the outer layers of skin cells (compare Figure 2.4). We have said that ionization is produced in biological tissue when the radiation transfers energy to it. If the α-particle only travels a short distance before losing all its energy, the resulting ionization is also produced over the same short distance. The particle is then said to be **densely ionizing**; as a result, any biological damage is concentrated in a few cells, or in a single cell, making it more difficult for them to repair the damage caused.

Because they are densely ionizing, α-particle emitters pose a serious radiation hazard if they enter the body—for example, if they enter the lungs. This is why radon-222, the gaseous radioisotope that forms part of the radioactive decay chain of ^{238}U (Figure 2.5), accounts for such a large fraction of the background radiation dose to the public (Figure 4.1). In some areas, radon formed from the decay of uranium in underlying rocks or building materials may infiltrate a building, and be inhaled. The inhabitants will then receive a dose of densely ionizing radiation when the radon or its daughter products decay in the lungs, which have no protective skin to provide any shielding.

Figure 2.4 showed that β-particles are more penetrating than α-particles: they have a range in air of several hundred millimetres. β-particles are high-energy electrons emitted from the nucleus during β-decay. Free electrons produced in other ways (for example by the interaction of γ-rays with matter, as we shall see shortly) will behave in the same way as β-particles. Because β-particles travel much further than α-particles, the ionization they produce is spread out over a greater distance; they are described as being **lightly ionizing**.

Unlike α-and β-particles, X-rays and γ-rays are uncharged, since they are both forms of electromagnetic radiation. Depending on their energy, X-rays and γ-rays are able to penetrate deep into tissue, and can even pass through the human body.

▷ How is this put to good use?

▶ In medical X-ray pictures. More X-rays are absorbed in the denser parts of the body, for example bone, than in tissue, enabling the different parts of the body to be identified on X-ray pictures.

However, when X- or γ-rays *do* interact with atoms, they do so by ejecting electrons, and it is these electrons that can cause ionization. Thus, the radiation effects of γ-rays and X-rays are those of lightly ionizing radiation.

Like X- and γ-rays, neutrons have no electric charge but are even more penetrating. Several centimetres of lead will protect you from high-energy X-rays or γ-rays, but, as you know from Figure 2.4 and Chapter 3, it may take a metre or more of concrete to provide an effective shield from a source of neutrons, such as a nuclear reactor. We saw in Section 2.3.2 that neutrons can interact with nuclei and cause charged particles to be emitted, normally protons or α-particles. It is these *charged particles* that cause biological damage from neutrons: both protons and α-particles are densely ionizing.

In terms of biological damage caused, it is the *density* of the ionization produced by each of these radiations along their path which is the most important factor. To describe this phenomenon quantitatively, radiations are characterized by their **linear energy transfer (LET)**, that is, by the energy that they transfer to the material they are traversing *per unit distance along their track.*

As living cells are typically of the order of a micrometre (10^{-6} m) in diameter, this is the characteristic distance used for measuring LET; it is normally quoted as the energy transferred (in keV) per micrometre. In this unit, lightly ionizing radiation would have an LET of a few keV μm^{-1} of tissue, corresponding to tens of ionizations per micrometre. On the other hand, densely ionizing radiation, like α-particles, would have a value of 100 keV μm^{-1} or more, producing thousands of ionizations per micrometre. The LET for protons lies between these two values, but they are classed as densely ionizing.

The passage of radiation through a material and the ensuing process of ionization take place very rapidly (in about 10^{-17} to 10^{-15} s). If ionization occurs in a material that has some biological significance—for example in DNA—then immediate cellular disruption may result. This is known as the **direct effect of radiation**.

However, the chance of an ionizing event taking place in DNA is not high because DNA represents only a small proportion of the total cell volume.

▷ What is the most abundant compound in a cell?

▶ Water; over 75% of the cell is water.

At first, it may seem that the ionization of a water molecule would not damage the cell as a whole. However, ionization of a water molecule can produce **radicals**. Radicals (denoted as X• for example) are electrically neutral atoms or molecules that have an unpaired electron in their outer electron shell (in stable atoms and molecules, electrons usually occur in pairs). This makes them chemically highly reactive.

When ionizing radiation interacts with a water molecule it can remove an electron to form a positive ion:

$$H_2O \longrightarrow H_2O^+ + e^- \tag{4.1}$$

This positive ion is unstable, and fragments as follows:

$$H_2O^+ \longrightarrow H^+ + OH• \tag{4.2}$$

The electron created in reaction 4.1 combines with H^+ to give a hydrogen atom:

$$H^+ + e^- \longrightarrow H• \tag{4.3}$$

Together, Equations 4.1–4.3 generate the radicals H• and OH•. These radicals may combine with atoms or molecules or with each other. For example, two OH• radicals may combine to form hydrogen peroxide:

$$OH• + OH• \longrightarrow H_2O_2 \tag{4.4}$$

which is chemically very reactive, being strongly oxidizing: peroxide blondes bleach their hair with it! Alternatively, the radicals may combine with protein molecules in solution, disrupting their original structure.

Even if the radicals are formed outside sites that are particularly sensitive to radiation, such as the cell nucleus, they can exist for long enough (up to 10^{-3} s) to diffuse into such target sites and interact chemically. Their stable reaction products, such as hydrogen peroxide, can diffuse even further. Therefore the consequences of ionizing one atom may be realized well away from the site of the original ionization—even in the next cell—leading to **secondary, or indirect, damage**.

Summary of Sections 4.1 and 4.2

1   Biological systems can be damaged by α-particles, β-particles, X-rays, γ-rays and neutrons, because these kinds of radiation cause ionization within cells.

2   Such ionization may break bonds in DNA or some other biologically significant materials; this is a direct effect of radiation. Secondary damage occurs when ionization subsequently leads to the formation and diffusion of radicals and their reaction products.

3   The larger the amount of energy absorbed in a radiation dose, the greater will be the ensuing biological damage. A short time between doses also leads to greater biological damage. The extent of the damage is dependent on the type of radiation.

4   α-particles transfer their energy within a short distance. So do neutrons, which produce charged particles by interacting with nuclei. These types of radiation have a high *linear energy transfer* (LET) and are densely ionizing, the biological damage being concentrated in a few cells or a single cell. The repair mechanisms in the cells cannot always then keep pace with the damage caused.

5   β-particles, and the electrons produced by X-rays and γ-rays, have a longer range than α-particles in biological materials. They have a lower LET than α-particles, and are lightly ionizing. The damage they cause can be more readily repaired than α-particle damage.

4.3 Units of radiation dose

Exposure to radiation may be due to radiation sources outside the body, or to radioactive substances that have been inhaled or ingested. In the latter cases, the effects of the resulting irradiation may depend on where the substance becomes located in the body (Box 4.2). For example, iodine in the body is concentrated in the thyroid gland, which may be damaged by any intake of radioactive iodine. In most cases, however, exposure to radiation tends to be a combination of both internal and external sources, and complex models are used to estimate the total radiation exposure received.

4.2

Box 4.2   Radiotoxicity

The risk from radioisotopes depends partly on what kind of radiations they emit, and partly on whether or not they become bound in the body. Radioisotopes that become bound in the body *and* produce intense biological damage when they decay are said to be **radiotoxic**, since even very small amounts can produce eventual death by inducing a cancer. For example plutonium is radiotoxic: it is an α-particle emitter, which accumulates in the liver and skeleton if it enters the bloodstream, and lodges in the lung if it is inhaled as dust.

Plutonium is not, however, 'the most toxic substance known to man' as is often claimed. This privilege probably falls on the biological toxins, like the ones that cause anthrax and diphtheria. Indeed, plutonium is only very poorly absorbed through the gut if it is ingested; most is excreted. ■

But for either internal or external irradiation, how is the radiation exposure quantified? As we saw earlier, it is the absorption of radiation energy in tissues, giving rise to ionization, which leads to biological damage. Therefore, one way of measuring the radiation exposure is to measure the energy absorbed in a unit mass of tissue. This is known as the **absorbed dose** and the SI unit is called a **gray (Gy)**.

1 gray = 1 joule of energy absorbed from radiation by 1 kg of tissue

However, as we have also seen, the amount of biological damage caused does not depend solely on the amount of energy absorbed. We have already discussed the fact that α-particles are biologically damaging because they are densely ionizing. Hence, for a given radiation dose—that is, energy absorbed per mass of tissue—the relative biological effect depends on the type of radiation involved.

In order to compare the biological effects of different types of radiation, the effect of different doses of the same type of radiation first has to be determined. Similar samples of cultured cells are exposed to different doses of the same type of radiation, and the number of surviving cells counted. Figure 4.4 shows a resulting **cell survival curve**. For the lower doses most of the cells survive because, when only small doses are given, the cells are able to repair any radiation damage caused. However, at higher doses some cells are unable to repair themselves, and the fraction of cells killed then starts to increase as the dose increases.

▷   Using Figure 4.4, work out the radiation dose, in grays, needed to kill 63% of the cell population?

▶   Note that the vertical axis plots percentage of cells *surviving*, so read off the radiation dose corresponding to 37% survival. This is about 2.3 Gy.

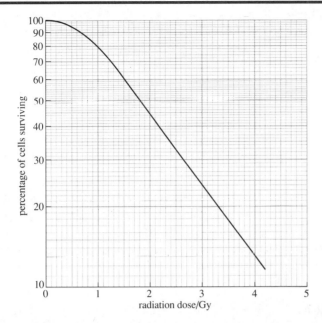

Figure 4.4 A cell survival curve. Note the logarithmic scale on the vertical axis. The radiation doses are all for a single type of radiation.

The symbol D_0 is often given to the absorbed dose needed to kill 63% of the cells. If these cell survival curves are repeated for different types of radiation, the **relative biological effectiveness (RBE)** of each type can be determined. The RBE for cell killing for a particular radiation is defined as:

$$\text{RBE} = \frac{D_0 \text{ for a reference radiation}}{D_0 \text{ for the radiation concerned}} \qquad (4.5)$$

The reference radiation is normally X-rays with an energy of 100 keV; in other words, X-rays with this energy are, by convention, assigned an RBE of 1.

▷ For a particular cell type, when the radiation is 100 keV X-rays, the absorbed dose, D_0 is 2.8 Gy, and it is 0.2 Gy using 5 MeV α-particles. What is the RBE for the α-particles?

▸ Using Equation 4.5, the RBE for the α-particles is:

$$\text{RBE (5 MeV α-particles)} = \frac{D_0 \text{ for 100 keV X-rays}}{D_0 \text{ for 5 MeV α-particles}} = \frac{2.8}{0.2} = 14.0$$

However, the radiation dose needed to kill 63% of a cell population—that is, D_0 —depends on the type of cell concerned. Some cells like those in bone marrow are much more easily killed by radiation than others, and humans comprise a whole range of different cell types, each with a different RBE for each different type and energy of radiation. In order to take this into account, a **quality factor (QF)** is determined for each type of radiation. The biological dangers of a particular radiation dose can then be expressed as a **dose equivalent**, which is:

dose equivalent = absorbed dose × quality factor   (4.6)

The dose equivalent is measured in the unit **sievert (Sv)**. Thus, 1 Sv corresponds to a dose of 1 Gy being given by radiation with a quality factor of 1. So the dose equivalent is a measure of the potential for biological damage from a particular absorbed dose of a particular type of radiation. In the same way that radiations differ in the biological damage that they cause, so the body organs differ in their response to the damage. For example, the lungs are more sensitive than the liver. To allow for this, the **effective dose equivalent** (also measured in sieverts) is used. This quantity weights the dose equivalent received by each organ with a factor that takes into account its sensitivity to radiation. The effective dose equivalent is often referred to simply as the **dose**, a practice that we shall follow hereafter. Where it is necessary to distinguish this from the absorbed dose (Gy), it will be clear from the unit used.

The QF values currently in use are shown in Table 4.1.

Table 4.1 Average quality factors for different types of radiation.

Radiation type	QF/Sv Gy^{-1}
γ-rays, X-rays and electrons (e.g. β-particles)	1
thermal neutrons	2–3
fast neutrons	10
α-particles	20

When evaluating the risks from radiation to a *group* of people, the **collective effective dose equivalent** is often used. This is simply the sum of the effective dose equivalents (in Sv) to individual members of the group, and is measured in the unit **man sievert (man Sv)**. If the collective dose is over a particular period, this is also specified—for example, man sieverts per year (man Sv yr^{-1}).

▷ The annual effective dose equivalents to four people living in an isolated community are 0.005, 0.002, 0.000 9 and 0.007 Sv. What is the collective annual effective dose equivalent for that population?

▶ The collective annual effective dose equivalent for that population is the sum of their individual effective dose equivalents. It is thus:

$$(0.005 + 0.002 + 0.000\,9 + 0.007)\,\text{man Sv yr}^{-1} = 0.014\,9\,\text{man Sv yr}^{-1}$$

Summary of Section 4.3

1 One measure of radiation dose is the energy that the tissue absorbs from the radiation. It is expressed in grays (joules per kilogram of tissue) and is known as the *absorbed dose*.

2 Densely ionizing radiations like α-particles are more effective in killing cells than lightly ionizing radiations like β-particles and γ-rays; this is taken account of by the *relative biological effectiveness*, RBE. In estimating the effects of radiation on humans, the different types of radiation are also assigned a *quality factor*.

3 A second measure of radiation dose, the *dose equivalent*, recognizes this varying biological effectiveness of the different types of radiation. It is the product of the absorbed dose and the quality factor, and is expressed in sieverts.

4 In order to take into account the varying sensitivity of body organs to radiation, the *effective dose equivalent* (measured in sieverts) is used. In practice, this quantity is often simply referred to as *dose*, which should be distinguished from the absorbed dose, measured in grays.

Question 4.1 Are the following statements true or false?

(a) The absorbed dose of radiation is a measure of how much energy is absorbed by a mass of tissue.

(b) The relative biological effectiveness compares the biological effectiveness of different radiations in producing a particular effect, for example killing cells.

(c) The use of the dose equivalent makes no allowance for the difference in biological effectiveness of different types of radiation.

(d) The collective effective dose equivalent is the sum of the individual radiation effective dose equivalents for a selected population.

4.4 The effects of radiation damage

The time taken for the biological effects of radiation to become apparent in a whole organism varies greatly. If the dose is high—greater than 20 Sv or so for humans— then the molecular damage is likely to be considerable. Observable effects begin to appear within minutes, hours or days of the exposure, as cellular disruption and death lead to whole organ failures and system degeneration. Such symptoms, and others that occur at levels above 2 Sv, are well documented, or fairly predictable in nature, and are known as **non-stochastic effects**. (The word *stochastic* literally means 'random', but in the context of ionizing radiation damage it would be more accurately defined as meaning 'unpredictable'.)

Stochastic effects are associated with low levels of radiation exposure. In keeping with the fact that they are unpredictable, they may not occur at all. If any effects do occur, they may not be observed until years, or decades, after the radiation exposure has taken place. Hence it is often difficult to establish a causal relationship between radiation exposure and stochastic effects.

An analogous example occurs in homeopathic medicine. In large doses, the effects of taking arsenic or extracts of belladonna (deadly nightshade) are well known.

▷   Would these effects be stochastic or non-stochastic?

▶   Non-stochastic. We can accurately predict what will happen: taking a large dose results in death or serious illness.

However, homeopathic practitioners dilute such compounds to an extreme degree and use them to treat some long-term complaints, such as asthma or arthritis. Here the outcome is not so certain and the effects, if they do occur, take place some time after the treatment has started. Hence an improvement in condition may be due to other factors (such as conventional medicine, or even natural remission). Many people are convinced that such remedies are effective, but the links between the treatment and its effects are often difficult to establish, leaving others sceptical. Any effects that do occur would be classed as stochastic effects.

The stochastic effects of radiation can be divided into two groups, the somatic effects and the genetic effects. Radiation damage, leading to a mutation at the level of either the gene or the chromosome, may result in cellular disorders, such as the uncontrolled growth of a cancer in the irradiated person. This is known as the **somatic stochastic effect**. However, if the radiation-induced mutation occurs in the reproductive cells of the body (gametes), the effects may appear not in the irradiated individual, but in any offspring the individual may subsequently have. This is known as a **genetic stochastic effect** (compare the distinction between somatic cells and gametes in Box 4.1). The genetic and somatic stochastic effects of radiation are the effects that may be seen at the low levels of radiation exposure experienced by us all. We shall consider them again in Section 4.7.

4.5 Assessing the evidence for determining radiation exposure limits

Defining 'safe' exposure levels for any poisonous or dangerous substance is very difficult. Firstly, what do we mean by 'safe'? Does this mean carrying no risk at all? If so, then all human activities would be unsafe. What one has to do, therefore, is to say what level of risk is 'acceptable', a concept which is fraught with problems, not least that the standards of safety that we expect appear to rise continually.

Having settled on an acceptable risk, we must then ask what level of radiation exposure corresponds to the risk that we are prepared to accept. Here, too, the question is difficult to answer because, as we shall see, getting evidence for the stochastic effects of radiation is a long drawn out and difficult process. Let us start by looking at some reputable estimates of the effects and risks that people experience when they receive different radiation doses.

Table 4.2 gives a summary of the radiation exposure levels, in terms of the instantaneous dose, which might be expected to result in observable, non-stochastic, effects. In addition, Table 4.2 also includes estimates of the likelihood of developing fatal cancer at the lower levels of radiation exposure. The latter are stochastic effects, and are quoted as **risks**, expressed numerically as one in some number or other. Box 4.3 clarifies the meaning of these risks.

Table 4.2   The effects of a single, instantaneous exposure to ionizing radiation on humans. The figures are those recommended by the International Commission on Radiological Protection (ICRP). A more recent, but not as yet completely approved, revision is discussed in Section 4.6.

Dose *	Effect
Non-stochastic effects (higher doses)	
> 20 Sv	central nervous system form of radiation sickness†
> 10 Sv	survival very unlikely
> 5 Sv	gastro-intestinal form of radiation sickness; damage to the lens of the eye
> 2 Sv	skin burns; bone marrow form of radiation sickness
0.1–1.0 Sv	temporarily low blood count; temporary sterility
Stochastic effects (lower doses)	
0.1 Sv (100 mSv)	estimated risk of fatal cancer 1 in 800
0.01 Sv (10 mSv)	estimated risk of fatal cancer 1 in 8 000 (which is roughly the same as the annual risk of being killed in a road accident in the UK)
0.001 Sv (1 mSv)	estimated risk of fatal cancer 1 in 80 000

* These values should be compared with the natural background radiation in the UK of just under 2 mSv per year.

† For example, this particular form of radiation sickness includes severe nausea and vomiting, loss of coordination, diarrhoea, siezure, coma and finally death.

4.3

Box 4.3   Risks

One entry in Table 4.2 states that a dose of 0.1 Sv is associated with an estimated risk of 1 in 800 of contracting a fatal cancer. You can think of this as an estimate of the outcome of an imaginary experiment. Suppose we take two equally large representative samples of the general population. By representative, we mean typical in age, sex and all other respects. We then subject each member of *one* of the samples to a dose of 0.1 Sv; the members of the other sample receive no treatment.

We now study the subsequent fate of the two samples. Then, according to Table 4.2, for every 800 people in each sample, there will be one more cancer death in the irradiated sample than in the unirradiated one. Suppose, for example, that each sample contains 800 000 people, and that in the unirradiated sample, 80 000 people subsequently die of cancer. In the irradiated sample, this figure will be raised by (800 000 × 1/800) or 1 000; that is, the cancer deaths in the irradiated sample will number 81 000.

Notice that the risks cited in Table 4.2 are those of *excess* cancers. Cancer is a very common disease, as our figure of 80 000 in the unirradiated sample implies. Abnormally high radiation doses only add to an already substantial risk of death from cancer.

Finally, note that there are other ways of expressing risks than that used in this Box.

Fractions, decimals or percentages may be used. Thus a risk of 1 in 100 could also be written as a risk of 1/100, or of 0.01, or of 1%. ■

You should now do Activity 4.2.

Activity 4.2

Activity 4.1 looked at the average annual radiation exposure for people living in the UK.

Use the data in Table 4.2 to work out your own risk of developing a radiation-related illness following one year's average radiation exposure, and say what that illness would be? What is the limitation of using data from Table 4.2 to estimate the risk from your *annual* radiation exposure (look carefully at the title of Table 4.2).

less than 1/80 000

So the risk from exposure to the average annual radiation dose in the UK is very small, and is classed as a stochastic effect. Indeed, even for doses up to about 1 Sv, some of the effects are stochastic, occurring many years later, with all the uncertainties and difficulties of defining cause and effect that this entails.

There are two main sources of information on which to base estimates of the risk of damage to humans from ionizing radiation. The first comes from animal experiments. The second consists of statistical studies of the effects in populations that have been exposed to radiation at doses and dose rates considerably in excess of those experienced either by workers exposed to radiation or by the general public.

▷ What categories of people have received a high dose of radiation?

▶ Survivors of atomic bomb explosions and people who have received medical radiation treatment are two such groups.

We shall now look at the data from these two populations, and discuss the difficulties in obtaining estimates such as those in Table 4.2 from them. We shall also look at the data from radiation workers, although they normally have much lower radiation doses, and lower dose rates, than the other two groups.

4.5.1 Patients receiving radiation treatment

Here, our concern is with people receiving radiation treatment for non-cancerous conditions, the largest and most important group of patients being those suffering from a stiffening of the spine, known as **ankylosing spondylitis.** During the period 1935 to 1954, 14 106 patients suffering from this disease were given X-ray therapy. High doses of radiation to the spine were successful in slowing the disease, but a small number of patients were subsequently found to develop cancer. This group has been closely followed, and the data obtained have been used to estimate the risk of developing cancer following radiation exposure. 18 patients developed leukaemia and another 70 developed other cancers in the heavily irradiated areas of the body.

▷ What percentage of the ankylosing spondylitis patients developed leukaemia?

▶ The percentage developing leukaemia is (18/14 106) × 100% = 0.13%.

The number of cancers including leukaemias (a total of 88) is more than one would expect in this group, based on the average risk of contracting cancer in the general population. The inference is that the incidence of cancer was a direct result of the radiation exposure that the patients received in their treatment. However, when the incidence of cancers, including leukaemia, is considered over a long period of time after radiation treatment, the number of *extra* cancers (compared to an unexposed population) decreases, until no more cancers are found in the exposed group than would normally be expected. Thus, it would appear that some subjects developed cancer *prematurely* as a result of their radiation exposure, but that there was no overall increase in the incidence of cancer.

4.5.2 Survivors of the atomic bombs dropped on Hiroshima and Nagasaki

In 1945, at the end of the Second World War, atomic bombs were dropped on Hiroshima and Nagasaki in Japan. The people in these cities were exposed not only to the force of the blast, and to associated fires, but also to an intense burst of radiation. About half the population of these cities died within five years from all the causes related to the nuclear explosions.

▷ A nuclear weapon is a supercritical assembly of fissile material. What types of radiation will be given off when it explodes?

▶ The radiations given off will be those from the fission reactions, namely neutrons, γ-rays and β-particles (see Figure 2.8).

▷ Which of these types of radiation are the most penetrating?

▶ Neutrons and γ-rays are the most penetrating (see Figure 2.4).

The main radiation dose to these populations came from the γ-rays, with a small fraction from the neutrons. Those who survived are by far the largest group to be studied for the effects of radiation exposure. One of the most important, and extremely difficult, tasks has been to estimate the radiation exposure of the people concerned. These estimates are still being revised, particularly in relation to the behaviour of the bomb itself, and to the weather conditions at the time.

Activity 4.3

Over 100 000 atomic bomb survivors have been monitored since the blasts. Of the 54 000 people in this group who incurred a significant absorbed dose of radiation (that is over 5 mGy), 3 832 had died of cancer between 1950 and 1982. In a normal population of the same size not exposed to this extra radiation dose, 3 601 cancer deaths would have been expected over the same period of time.

(a) From the data given above, calculate what percentage of the population who received over 5 mGy had died from cancer during the period of study.

(b) Now calculate what the percentage would have been if the atomic bombs had not been dropped on Hiroshima and Nagasaki.

(c) What is the percentage increase in cancer deaths that could be attributed to the radiation exposure experienced due to the atomic bomb explosion? What is this as a percentage of the whole population?

Activity 4.3 illustrates two points already touched on in Box 4.3. Firstly, cancer is a fairly common disease. Without radiation exposure we would have expected nearly 7% of the 54 000 people to have died from cancer during those 32 years anyway. Secondly, although the people in Hiroshima and Nagasaki were exposed to quite high levels of radiation, only a small number (231) of extra fatal cancers were induced above the 3 601 to be expected. In this case, the excess cancers are numerous enough to be attributable to the radiation from the atomic bombs. However, when the dose rate is much lower, the excess deaths from cancer will be a much smaller proportion of the normal cancer death rate. It will then be difficult to prove that they were caused by the incident, rather than by natural variations in the normal death rate from cancer over the same period. This difficulty would be encountered, for example, when assessing the effect of the Chernobyl accident on cancer rates in Europe (Section 6.4.5).

At Hiroshima and Nagasaki, leukaemia was the first cancer to appear in greater numbers than would otherwise be expected, and this excess was seen for about 20 years following the exposure. However, as with the ankylosing spondylitis patients, this excess diminished as the study period increased. Most of the other cancers seen, however, take longer to develop than leukaemia, and there continues to be an increased incidence of these cancers.

The frequency of these radiation-induced cancers in the Hiroshima and Nagasaki survivors will continue to be monitored in future years, and should lead to an answer to a key question: will they increase in frequency as the population that was exposed to the atomic bombs ages? This is the normal pattern for naturally occurring cancers, as Figure 4.5 shows. The question that can then only be answered when all the atomic bomb survivors have died is, 'Is the overall incidence of cancer greater among the bomb survivors than in the normal population?'. Because many of the survivors were children at the time of the bomb explosions, the answer to this may not be known until 2020 or later.

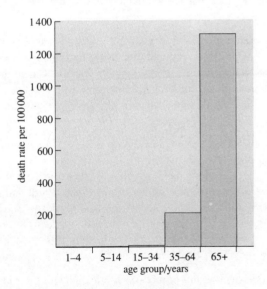

Figure 4.5 Incidence of cancer deaths by age in England and Wales (1986–90).

4.5.3 Occupationally exposed workers

Four large-scale studies have been undertaken to assess the risk of radiation-induced disease in occupationally exposed workers, by looking at the groups of people who work in nuclear power plants. One of these studies was of workers at the Hanford plant in Washington State, USA; the second was of 10 000 employees of British Nuclear Fuels (BNFL); the third was of 40 000 United Kingdom Atomic Energy Authority (UKAEA) workers; the fourth was of 95 000 workers on the UK National Registry for Radiation Workers.

These studies involved relatively low-level radiation exposures and were designed to detect the stochastic effects that concern us. All four, however, demonstrated what is known as the **healthy worker effect**; that is, deaths from all causes, and not just deaths from cancers, were found to be *lower* in the three study groups than in the general public.

▷ Why do you think that workers in these plants should be healthier than the general public?

▶ There are two main reasons. Firstly, the workers concerned must by definition be 'fit for work', whereas this is not true of the whole population, even when the different age structure is taken into account. Secondly, radiation workers undergo regular medical check-ups, where diseases can be diagnosed in their early stages and successfully treated. As yet, this benefit is not widely available for the rest of the British and US populations through their health services.

None of the first three studies was able to establish clear evidence of a relationship between the cancers found in the radiation worker populations and their low levels of radiation exposure. However, the fourth study did establish such a relationship, but considerable controversy still remains.

4.5.4 The dose–response curve

The radiation doses received by the atomic bomb survivors, and radiation-treatment patients, were, in the main, at least fifty times the natural background of just under 2 mSv per year (Activity 4.1). So, to estimate the risk associated with low radiation levels we have to use the data that are available from higher radiation exposures, and to extrapolate from them.

Figure 4.6 shows an example of what is known as a **dose–response curve**. Here the number of excess cancer deaths in a population is plotted against the radiation dose received by that population. The data points are scattered, and it is not possible to draw a smooth curve through all the points. How then do we extrapolate these data to low doses?

It is often assumed that, no matter how small the radiation dose is, it is always associated with *some* increased risk of developing cancer. If this is so, then the curve through the data should go through the origin. If we assume this, and it is by no means a universally accepted view, then we have to predict how that curve behaves towards the lower end of the exposure range, between the origin and the data points. One method of extrapolating from the effects seen at high doses to those to be expected at low doses would be to assume that dose and effect are directly *proportional*. This means that the data should fit a straight line; this is called the **linear model**, and is shown on Figure 4.6.

An alternative model, known as the **quadratic model**, postulates that the additional risk of developing cancer as a result of the radiation dose is proportional to the *square* of the radiation dose. The basis for this model is the idea that, in order for a cell to become cancerous, the DNA strand (or other target site) has to be broken in two places at once. It has been shown by experiment that a single break in a DNA strand can be repaired. Therefore, at low levels of radiation exposure, where only single breaks are likely, the chances of a cancer developing are reduced by this repair mechanism. However, at higher exposures there is an increased chance of the target site being damaged in two places simultaneously, and, according to the quadratic model, the repair mechanism fails.

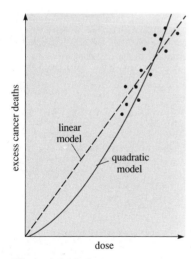

Figure 4.6 A dose–response curve. No units are given because this plot is only to illustrate the principles discussed.

Despite many animal and cell-culture studies, it remains uncertain which of the two models is the more reliable.

▷ According to Figure 4.6, will there be an over- or underestimate if the linear model is used to predict the risk of radiation-induced cancer at low exposures when a quadratic model is the correct one?

▶ If the linear model is used, but the quadratic model is correct, then the risk at low levels will be overestimated.

It has even been suggested that, far from being harmful, low levels of radiation may actually be *beneficial*, because a low-to-moderate radiation exposure may stimulate the body's immune system. This, in turn, may improve the ability of the immune system to recognize and destroy mutated and misrepaired cells. This phenomenon of **radiation hormesis** is proposed to explain the apparent lack of excess cancers seen in the lower-exposure categories of the atomic bomb survivors; it would also explain the fact that in some animal experiments the animals live longer than normal following low radiation doses. Figure 4.7 shows the results of such an experiment. At a dose rate of 0.01 Gy week^{-1}, the average survival time is greater than 100%, so these mice seem to be living longer than the non-irradiated controls.

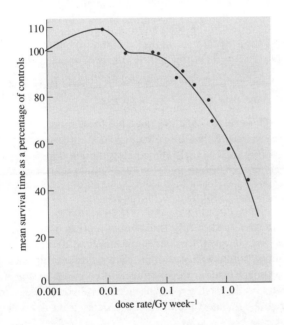

Figure 4.7 Average survival times for mice continuously irradiated with γ-rays as a percentage of those for non-irradiated animals, plotted against the dose rate. Note that dose rate is plotted on a logarithmic scale.

▷ Why, on the basis of Figure 4.7, should you be sceptical about this conclusion?

▶ Only a single point in the Figure is above 100%. More points would be needed to be confident of the effect, or to establish a 'threshold' radiation level below which increased life expectancy became significant.

However, this is not an isolated example, and indeed, the effect has been observed more often than it has been reported, because those observing it often felt that there must be something wrong with their data! (Scientists are human, and are often reluctant to reveal what they believe to be their own mistakes.) Nevertheless, like many of the effects of radiation, the hypothesis that radiation hormesis does occur requires much more investigation.

Summary of Sections 4.4 and 4.5

1   High instantaneous radiation exposures of over 100 mSv have non-stochastic effects. Low exposures have stochastic effects that are expressed as *probabilities*, notably those of contracting fatal cancers.

2   Data on such effects have been obtained mainly from studies of excess cancers among survivors of atomic bomb explosions, and patients who have received high doses during radiation treatment.

3   Risks of stochastic effects at low exposures can be obtained by extrapolating the data referred to in point 2 above to low radiation levels using either a linear or quadratic model.

4   There is some evidence that low levels of radiation may be beneficial.

4.6 Setting radiation dose limits

An international body, the International Commission on Radiological Protection (ICRP), reviews the data and makes recommendations on radiation dose limits. In the UK, the National Radiological Protection Board (NRPB) is responsible for advising the government on what the radiation dose limits should be. The two basic principles that the ICRP applies to radiological protection are as follows:

1   No practice involving exposure to radiation should be adopted unless it produces a net benefit to the population (how this net benefit is assessed is not specified).

2   All exposures to radiation should be As Low As Reasonably Practicable (the ALARP principle).

These principles apply generally, and are intended to be more restrictive than dose limits, which are set to ensure that the risk to an individual is effectively controlled. Thus, those who oppose nuclear fuel reprocessing (which we shall discuss in Chapter 5) argue that there is no net benefit to be derived from undertaking it, and therefore that any resulting radiation exposure cannot be justified.

In order to arrive at 'safe' dose limits, some estimate of the risks from a particular radiation exposure has to be made. We have already given you a set of risk estimates of this sort in Table 4.2. It was provided by the ICRP. A more recent revision exists which has not yet been completely accepted. High doses are represented by instantaneous doses exceeding 100 mSv and capable of giving rise to non-stochastic effects; below this the effects are likely to be stochastic. This distinction is consistent with the one made in Table 4.2. The latest (1990) risk estimates from the ICRP are as follows.

1   *The risk of hereditary disease*: no radiation-induced hereditary diseases have been observed in humans. Therefore the estimate of a 1% risk per Sv of radiation dose received (that is, a 1 in 100 chance of the disease occurring if a dose of 1 Sv is given) is based on animal data alone.

2   *The risk of developing a fatal cancer following a high radiation dose* (> 100 mSv) is 10 in 100 (or 10%) per Sv of radiation dose received.

3   *The risk of developing a fatal cancer following a low-level radiation dose* is less, because of the ability of the DNA in cells to repair itself. In this case it is a 5 in 100 (or 5%) risk per Sv of radiation dose received. (*Note*   This risk is for a mixed population, including children, who are known to be more susceptible to the effects of radiation. In an adult population, such as a group of workers, the risk is reduced to a 4% risk per Sv of radiation dose received.)

4   For both radiation workers and the general public, *the risk of developing a non-fatal cancer following a low-level radiation dose* is 1 in 100 (or 1%) per Sv of radiation dose received.

The ICRP and other organizations have compared the risks associated with other types of activity, both in the home and at work, with the risks associated with various levels of radiation exposure using the above figures. Based on these comparisons, the ICRP has recommended that the radiation **dose limits** arising from human activities should be:

for the *public*, a maximum of $1\,\text{mSv}\,\text{yr}^{-1}$, averaged over 5 years;

for *radiation workers*, a maximum of $20\,\text{mSv}\,\text{yr}^{-1}$, averaged over 5 years.

The reason why the recommended doses to radiation workers is greater than that for the public is that the exposure of radiation workers is carefully monitored. The risks involved can therefore be controlled. In contrast, exposure of individual members of the general public is not monitored, so the likelihood of their receiving an unacceptably high dose by chance is reduced by lowering the radiation levels in areas where they might be exposed, for example in the vicinity of nuclear power stations or in hospitals that use radioactive sources.

Let us try to put these recommendations into perspective. We saw earlier that the natural background radiation in the UK gave an average radiation dose of just under $2\,\text{mSv}\,\text{yr}^{-1}$, so that this recommended limit for *additional* exposure of the public is about half what is normally received.

Turning now to radiation workers, suppose that a worker were to receive a dose of 20 mSv in a year (the maximum *average* amount recommended), what would be the risk of dying from a radiation-induced cancer using the ICRP's latest recommended risk estimates? From the distinction that we made earlier in this section, a high radiation dose would be an instantaneous dose exceeding 100 mSv. The worker's dose is only 20% of this, and is spread over a year, so that this dose would come in the 'low-level' category in the ICRP's risk estimates. The risk figure that we should use is that for an adult population. From the estimated risks of developing a fatal cancer given in point 3 above, the correct one to use would be 4% per Sv.

Using this value, a dose of 20 mSv (0.02 Sv) corresponds to a risk of (4% per Sv × 0.02 Sv), or 0.08%. The worker therefore has an estimated 0.08% chance (1 in 1 250) of dying from cancer as a result of such an exposure.

Had you used the data in Table 4.2 to estimate the risk of contracting a fatal cancer following exposure to a radiation dose of 20 mSv, you would have found that it was between 1 in 800 and 1 in 8 000, which is consistent with the estimate we have just made based on the more recent ICRP recommendations.

4.7 Leukaemia clusters around nuclear installations

As we saw earlier, cancer is a very common disease: 1 in 5 people in the UK die from some form of it. Conversely, leukaemia is relatively rare, affecting, only about 1 in 1 800 of the population. However, *childhood* leukaemia accounts for around 10% of all leukaemias, and kills more children between the ages of 2 and 15 than any other disease.

It is thought that leukaemia can begin with a disruption of the cells that give rise to blood cells in the bone marrow. The body's immune system fails to recognize that the cells are faulty, and they escape destruction by antibodies. We have already seen (for example, in Section 4.5.1) that high doses of ionizing radiation can lead to the onset of leukaemia. However, some chemical agents, notably benzene, and certain viruses, may also cause this type of cell damage.

In 1983, James Cutler, a journalist with Yorkshire Television, noticed that at Seascale, a small town near the Calder Hall nuclear reactors and the fuel reprocessing plant at Sellafield, there seemed to be more cases of childhood leukaemia than he expected—a so-called **leukaemia cluster**. Following the broadcast of a programme he made on this issue, a committee of inquiry was set up by the government. This committee reported its findings (in the 'Black Report' named after the chairman, Sir Douglas Black) in 1984, confirming the excess of childhood leukaemias in the area, but stating that more evidence was needed before any relationship between radiation exposure and the leukaemia cases could be confirmed.

Leukaemia clusters have been observed elsewhere in the country, many occurring in areas remote from any plant emitting radioisotopes. A number of theories have been proposed to explain their presence. One is that the clusters are due to a viral origin for leukaemia. If the occurrence of the virus is localized, then people living in a stable, isolated community will develop immunity against it. However, the immune protection will not be present in newcomers to the area—in particular those moving to a new town or migrant workers building a large factory nearby—and in such populations excess leukaemias are found. This has certainly been the case in the establishment of some new towns, such as Corby and Cwmbran for example, where excess leukaemias have also been observed. However, whether this is the explanation for all leukaemia clusters remains an open question.

In 1990, a report in the *British Medical Journal* by Professor Martin Gardner, of the MRC Epidemiology Unit at Southampton, suggested that there may be a possible link between radiation exposure to the fathers prior to the conception of the children and subsequent leukaemia in those children. Gardner reported a statistically significant association between paternal occupation and childhood leukaemia, whereby fathers working in the nuclear fuel reprocessing plant at Sellafield showed an increased risk of having a leukaemic child. Of the five leukaemia cases diagnosed in children born in the small area in and around Seascale, four had fathers known to have worked at the Sellafield plant, and the fifth was thought to have worked there, but could not be traced on the Sellafield computer file. Three of the fathers who had definitely worked at Sellafield were in the high-exposure group at the plant (having total radiation doses of greater than 100 mSv), and the fourth had also been subject to significant radiation exposure.

The question of the possible association between parental radiation exposure and the incidence of leukaemia is clearly an important one. It is also a particularly emotive issue, because children are involved. Why is it so difficult to establish the link? Extract 4.1 puts the findings in perspective and also indicates some of the problems of establishing the relationship. You should read this with the questions posed in Activity 4.4 in mind.

Activity 4.4   *You should spend up to 20 minutes on this activity.*

Notes relating to Extract 4.1.

1   'Controls' are people who are used as a reference population. They are people who have not been exposed to the suggested source of the risk, or who have not contracted the disease. Where increases in leukaemia, or relative risks are noted, this is with respect to the rate in the control population, either in the area (west Cumbria) or more locally.

2   When the text speaks of exposure to radiation being a *surrogate* for exposure to something else, it means that there may be some unknown danger in working at Sellafield, other than radiation, which is responsible for the excess childhood leukaemias.

3   Dosimeters are devices for measuring radiation dose, of which film badges are one type. Radionuclide is another word for radioisotope.

Read Extract 4.1 and answer the following questions.

(a) What were the radiation doses to the workers associated with a large increase in the risk of leukaemia in their children?

(b) By what factor was the risk of leukaemia in their children increased?

(c) In addition to the father's radiation exposure, what other possible causes of the leukaemias were investigated?

(d) What other evidence may be in conflict with the findings of the Gardner report, and why is it not directly comparable?

(e) What other factor needs investigating as a possible cause of childhood leukaemia in the offspring of exposed workers?

Extract 4.1   From *British Medical Journal*, 17 February 1990, pp. 411–12.

Leukaemia and nuclear installations

By Valerie Beral, Director, Imperial Cancer Research Fund, Cancer Epidemiology Unit, Radcliffe Infirmary, Oxford

Occupational exposure of fathers to radiation may be the explanation

The study of leukaemia and lymphoma in West Cumbria reported today (p. 423) was commissioned by the Black inquiry into the raised incidence of childhood leukaemia in the village of Seascale near the nuclear plant Sellafield. The risk of childhood leukaemia was found to be unrelated to various indices of environmental contamination from the Sellafield discharges, such as eating seafood or home grown vegetables or playing on the beach. But the risk was raised if the children's fathers had been employed at Sellafield, particularly if they had had relatively high exposures of radiation before the affected children were conceived. The numbers are small, but the effects are large. The fathers of nine (out of 46) cases and 41 (out of 277) local controls were working at Sellafield when the child was born; but four of the case fathers and three of the control fathers had accumulated exposures to 100 mSv or more of external radiation before the child was conceived. An exposure of 100 mSv or more was associated with a sixfold to eightfold increased risk of leukaemia in the offspring. There was evidence of an increased risk at exposures lower than 100 mSv only when data for the six months before conception were considered, but the numbers were exceedingly small…This study by Gardner and his colleagues is the first to examine the relation between paternal employment in the nuclear industry and the risk of leukaemia in the offspring. Some coments seem appropriate at this stage even though the children of other nuclear workers need to be studied before firm conclusions can be drawn. Three separate inquiries into alleged increases of childhood leukaemia near different nuclear installations each concluded that there was a real excess but that the increases were too large to be accounted for by radioactive discharges from the plants. Each report emphasised that alternative—but as yet unknown—pathways of exposure and mechanisms of carcinogenesis needed to be considered. The results of this study by Gardner *et al.* are remarkable not because they offer little support for environmental contamination by radioactive discharges being the cause of childhood leukaemia but because they point to possible alternatives.

According to Gardner *et al.*, fathers' employment at Sellafield is sufficient to account for the raised incidence of childhood leukaemia in the vicinity. Could paternal employment account for the raised incidence of childhood leukaemia near other plants? The relative risk of childhood leukaemia ranges from 1.4 near Aldermaston and Burghfield to 5 near Dounreay and 10 near Sellafield. This range is incompatible with the cause being environmental exposure to radiation: if that were the cause of the childhood leukaemia the relative risks would need to vary more than 1 000-fold, since the estimated annual exposure of newborn infants from radioactive discharges ranges from 0.000 01 mSv at Aldermaston to 0.005 mSv at Dounreay and 0.3 mSv at Sellafield. The variation in occupational exposure at the plants is much less: the average in radiation workers ranges from 7.8 mSv at Aldermaston to 47.0 mSv at Dounreay and 124.0 mSv at Sellafield. Thus the range of occupational exposure and the different mix of nuclear and other workers in the surrounding community is not inconsistent with the range of leukaemia risks observed.

The explanation offered by Gardner *et al.* is not, however, without its problems. The only other relevant human data available are on the 7 400 children of Japanese men who survived the atomic bomb explosions, and these show no hint of an increased risk of leukaemia in the offspring. And the average exposure to external ionising radiation of the Japanese men was four times higher than that of the Sellafield workers. Some additional explanation will still be required for the children of Sellafield workers. For example, it could be argued that exposure to high levels of radiation at work is a surrogate for exposure to something other than radiation which itself is powerfully leukaemogenic in the next generation. There is, however, no known substance which increases the risk of leukaemia in offspring of those exposed. It is also possible that the most heavily exposed workers might inadvertently bring radioactive materials home—for example on contaminated clothing. Some studies have found unusually high concentrations of some radionuclides in the dust of workers' homes, but the extent of this domestic contamination was probably not sufficient to explain the excess of childhood leukaemia near Sellafield. Another possibility is that internal rather than external radiation exposure is relevant. If workers were internally contaminated with a radionuclide which was concentrated in the urogenital organs or the semen, the doses to the germ cell or the fetus could be greater than those recorded on the worker's externally worn dosimeters or film badges. The risk of prostatic cancer has been shown to be increased in some of the most heavily exposed employees of the United Kingdom Atomic Energy Authority and of the Atomic Weapons Establishment, and it has been suggested that some radionuclides may be concentrated in the prostate. In both workforces the risk of prostatic cancer was increased more than 10-fold in the small group of workers with exposures to external radiation of 100 mSv or more, who also had been monitored for possible internal contamination by many different types of radionuclides, including tritium, plutonium, and uranium…

The results reported today are the first of their kind, and the risks described have large uncertainties associated with them. It would be premature to recommend formal changes to radiation protection limits on the basis of this one study; but until the findings from other studies are available workers need to be counselled and those who have not yet completed their families should be advised to avoid high exposures. The nuclear industry and its workforce have a good record for voluntarily limiting exposure and for collaborating with independent researchers in studying the health of the workers. This needs to continue. The more rapidly the mystery of childhood leukaemia near the nuclear plants can be unravelled the more rapidly steps can be taken to prevent it.

4.7.1 Recent developments in explaining the relationship between leukaemia and radiation exposure

In Extract 4.1, the fact that there has been no excess of leukaemias in the children of atomic bomb survivors was identified as contradicting the findings of Gardner and his colleagues. However, more recently, new work was reported which could explain this anomaly. As noted in Section 4.5.2, atomic bomb survivors were primarily exposed to an *external* radiation dose, the most penetrating part of which consisted of γ-rays plus a small proportion of neutrons. Now do Activity 4.5, which deals with this new work.

Activity 4.5

Read Extract 4.2 and then answer the following questions:

(a) How might the work described in Extract 4.2 account for the absence of excess leukaemias in the children of atomic bomb survivors, and its presence in the offspring of Sellafield workers?

(b) Do you agree with the claim made in Extract 4.2 that α-particles 'are superficially much less harmful than more penetrating radiation'?

Extract 4.2 From *The Times*, 20 February 1992.

Leukaemia clusters linked to low radiation

By Nigel Hawkes Science Editor

Levels of radiation believed to be too low to be harmful could in fact cause damage linked to leukaemia, according to Medical Research Council scientists. Their findings may explain the 'clusters' of the disease around nuclear sites such as Sellafield.

The research focuses on alpha rays, which are superficially much less harmful than more penetrating radiation. The scientists have found that alpha particles can cause hidden damage to cells that only becomes apparent some time later. The cell survives, and continues to divide apparently normally. Later, however, abnormal chromosomes appear in the successive generations, showing that the genetic material of the cell has been damaged.

The implication is that the lowest imaginable dose—a single alpha particle—is enough to induce damaging changes in some cells that can alter the cells' behaviour and perhaps lead to cancers. The scientists have found that X-rays do not produce the same effect.

Though reluctant to jump to conclusions, they believe that their findings may help to explain the incidence of leukaemia both among radiation workers and children living close to nuclear plants such as Sellafield.

The results could also prompt a change in the approach to radiation protection, which is based on the effects of high-energy penetrating radiation, such as beta and gamma rays.

Alpha rays appear less harmful, being unable even to pass through a sheet of paper. Their danger arises when particles of alpha-emitting isotopes such as plutonium are swallowed or get into the lungs. Then the alpha particles can reach living cells, unloading all their energy in a single hit and doing considerable damage.

Eric Wright and colleagues from the research council's Radiobiology Unit in Didcot, Oxfordshire, report in this week's *Nature* that they have studied the effect of single alpha particles on stem cells from the bone marrow of mice. In mammals, these stem cells have the job of producing all the blood cells, so any damage to them could well lead to blood diseases such as leukaemia.

When the stem cells are exposed to alpha particles, most are killed outright, but a small percentage survive apparently uninjured. Several generations later, when the cells have divided repeatedly, gross abnormalities can be detected in their chromosomes. The research council says that the experiments show that there are types of damage unique to alpha-particle radiation.

Dr Wright said yesterday: 'The mouse stem cells are a good model for leukaemia, and now we are trying the same experiment with human bone marrow cells.'

Clearly, we now have the beginnings of a possible explanation of how childhood leukaemia can be linked to radiation exposure, and there may soon be enough pieces of the jigsaw for the true picture to emerge.

Summary of Sections 4.6 and 4.7

1 The most recent ICRP recommendations for radiation dose limits in excess of natural background are, for the public, $1 \, \mathrm{mSv \, yr^{-1}}$ averaged over five years and, for radiation workers, $20 \, \mathrm{mSv \, yr^{-1}}$ averaged over five years. The increased risk of a fatal cancer to a worker receiving a $20 \, \mathrm{mSv}$ dose is estimated to be about 1 in 1 250.

2 One proposed explanation of some of the leukaemia clusters that occur in the UK is that the immunity of isolated communities to a leukaemia virus is lost when a new town or factory brings an influx of new people.

3 One leukaemia cluster occurs near the Sellafield nuclear complex. There appears to be a significant association between childhood leukaemia and paternal employment at the nuclear reprocessing plant, the leukaemia risk being greater when the father's radiation dose has been larger.

4 The importance of internal radiation doses from α-emitters in producing biological damage might explain why there has been no excess of leukaemias among the children of atomic bomb survivors, who were not exposed to α-particles.

external radiation
internal "

5 The nuclear fuel cycle

The nuclear fuel cycle starts with the extraction of uranium from the Earth's crust, and ends with the disposal of nuclear waste back into the crust. In this chapter we shall give brief details of some of the industrial processes involved, in order to give you a feeling for the scale and range of the activities. At each stage, we shall identify the aspects that give rise to risks, either to workers in the industry or to the general population. For the general population, these risks are all associated with the release of radioisotopes into the environment.

There are five distinct stages in the nuclear fuel cycle (Figure 5.1). The first is the mining of uranium ore, and the extraction from it of the uranium compounds in an impure but concentrated form.

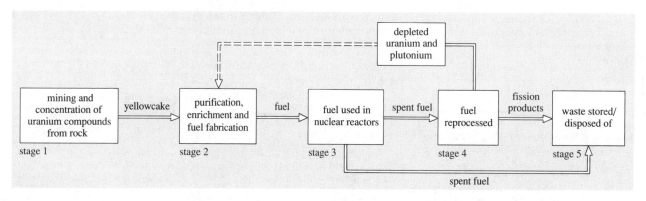

Figure 5.1   The nuclear fuel cycle.

In order to make nuclear fuel, it is necessary to convert the uranium concentrate into the right chemical form. It may also be necessary to change its isotopic composition. These processes, together with the fabrication of fuel elements, form the second stage of the cycle. The third stage is the use of the fuel in a nuclear reactor. After the fuel has supplied energy in the reactor, it is removed. The spent fuel, which now contains plutonium and fission products as well as ^{238}U and unused ^{235}U, can be dealt with in essentially two ways. In one of these, the fuel is reprocessed in stage 4 of Figure 5.1. This means that the uranium, plutonium and fission products are separated from each other. Of these three, now separate, components, the fission products are by far the most radioactive, and they must be stored and ultimately disposed of (stage 5 of Figure 5.1). The unused uranium and plutonium can, in principle, be returned to nuclear reactors to provide more fuel, a process indicated by the broken arrow between stage 4 and stage 2 in Figure 5.1. However, for reasons that we shall discuss in Section 5.8.2, this option has been little used up until now, and nuclear fuel reprocessing has therefore led to an accumulating stock of plutonium and uranium. If this uranium has a ^{235}U content less than the 0.72% found in natural uranium, it is said to be **depleted**.

The accumulation of plutonium has raised questions about the desirability of reprocessing and it explains the second way of dealing with spent fuel. Here the reprocessing stage 4 is bypassed, and the fuel rods containing uranium, plutonium and fission products are treated as waste to be stored and ultimately disposed of.

5.1 Mining uranium

Uranium is one of the rarer elements in the Earth's crust, with an average abundance of about 2 parts per million by mass. This, however, still makes it nearly as plentiful as tin, and roughly 20 times as common as silver. Uranium mining is undertaken at sites where the element has been concentrated by geological processes. In some places, the concentration approaches 4% by mass; at present, extraction is economically profitable if the concentration exceeds about 0.04%.

Figure 5.2 shows that the distribution of uranium deposits is far from uniform worldwide. The mining of uranium occurs both on the surface (principally for large, low-concentration deposits) and underground (for veins of high-concentration material).

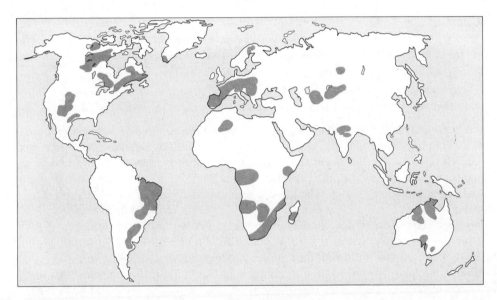

Figure 5.2 The distribution of the world's known major uranium deposits.

Uranium usually occurs as its oxide, uraninite (approximate composition U_3O_8), also known as **pitchblende**. However, since the uranium concentrations found are generally low, a large amount of waste material is produced when the uranium is extracted from the rock. For example, in a high-concentration ore, there might be 1% of uranium oxides; this means that 1 t of ore will contain 10 kg of uranium oxides and 990 kg of what is potentially waste. Extraction involves crushing the rock and treating it chemically to separate out the uranium; this is usually undertaken at the mine. The uranium leaves the mine as **yellowcake**, an impure form of uranium trioxide, UO_3; this is sometimes called uranium **concentrate**.

5.1.1 The risks from uranium mining and extraction

We saw in Chapter 2 that both ^{235}U and ^{238}U are radioactive (albeit with very long half-lives), and Figure 2.5 showed the radioactive decay chain for ^{238}U; there is a similar one for ^{235}U, which ends in the stable isotope ^{207}Pb.

All the isotopes in the decay chains of both ^{235}U and ^{238}U will be present in the uranium ore; because they are chemically different from uranium, they will usually form part of the waste material. Looking back to Figure 2.5, you can see that one of these isotopes (radium-226) is the precursor of the gaseous isotope radon-222, which will be released from the waste material.

▷   Why does radon-222 pose a particular health hazard?

▶   Because it is a gas, and can therefore be inhaled. If it undergoes radioactive decay when it is in the lungs, the α-particle emitted will cause damage to the lung surface. In addition, the daughter products are non-gaseous, and will lodge in the lungs, increasing the total radiation dose when they decay (Section 4.2).

You may wonder why this release of radon, or its precursors, following mining is important, because the production of the radon would have taken place in the ore *anyway* when it was underground. But if the ore is deep underground, the radon could take several days to diffuse to the surface. As its half-life is only 3.8 days, much of it will have decayed into the non-gaseous radioisotope ^{218}Po, which will not continue diffusing to the surface. The chances of breathing in radon-222 will therefore have been reduced.

Question 5.1 Radon-222 (half-life 3.8 days) from a deep underground mine takes 7.6 days to diffuse to the surface. What percentage of the original activity will be left by the time that it reaches the surface? (*Hint* You can use Equation 2.7.)

This example demonstrates the extent to which the radiation hazard associated with the decay of radon-222 will be reduced if it has to diffuse through overlying rock.

In open-cast mines, any radon produced is rapidly dispersed into the atmosphere, and does not pose too great a risk to the workers, but in underground mines the radon gas tends to accumulate. 'Mountain sickness' was the name given by the early uranium miners to the lung diseases they developed from breathing in both radioactive dust and, most importantly, radon. The rapid rise in uranium mining in the USA following the 1946 Atomic Energy Act resulted in very significant radiation exposures to those involved. The effect of these exposures on the miners has been documented by the Union of Concerned Scientists (*The Nuclear Fuel Cycle*, MIT Press, 1975, pp. 20–33). They estimated that the risk of lung cancer is doubled for every 2.4 Gy of radiation dose to the lungs from radon and its daughter products. They also estimated that the mean lung dose to the miners was 7–10 Gy, and that the maximum could be as high as 100 Gy.

▷   If the risk of lung cancer increases by a factor of two for every 2.4 Gy lung dose, by how much is the risk of lung cancer increased for miners whose lung dose is 7.2 Gy.

▶   A lung dose of 7.2 Gy is a factor of (7.2 Gy/2.4 Gy) larger than the dose that doubles the risk—that is, a factor of three higher. So, if 2.4 Gy increases the risk by a factor of two, three times that will produce (2 × 3)—that is, six times the risk.

With the maximum dose of 100 Gy, similar calculations show that the risk of lung cancer increases by a factor of 83. Substantial though these risks are, the problem of exposure can be readily solved. Thus, the Union of Concerned Scientists says simply that (p. 31) 'radon and its daughters can be removed from mines by ventilation'. They suggest that the radon levels should be reduced to a level such that, over 25 years, the total lung dose from radon and its daughter products is less than 0.1 Gy. This, they estimated to be one-tenth of the dose needed to induce a lung cancer. They further estimated that the cost of reducing the radon level in this way would be between 10 and 20% of the value of the uranium mined.

The cost of uranium is only a tiny fraction of the total cost of nuclear power, so that a 20% increase in its price presents no problems in principle. In practice it may mean that the mines covered by such standards become less competitive (Box 5.1).

Box 5.1   *Political and ethical issues*

At various points in this Course, some of the issues involve *value judgements*—political or ethical issues on which opinions can be very varied and often extreme. Sometimes, apparently straightforward decisions can raise important issues of principle. Uranium mining is one example.

▷   What could be one consequence of always buying uranium from the cheapest source?

▶   It could discourage investment in safety because the cost of increasing safety is a factor in determining the price.

The industrialized, and richer countries often invest in automation in order to reduce the risks associated with different production processes, and to make them more efficient. This usually also decreases the labour requirements, and hence the number of workers at risk. However, workers are the main asset of the poorer countries, and these countries do

not have the wealth to invest in automation. Consequently, if they are required to meet the same safety standards as those in industrialized countries and, at the same time, continue to use a large labour force rather than invest in automation, they could run the risk of becoming uncompetitive, and hence of losing trade.

Put more starkly, safety is a luxury for those whose alternative to unsafe employment is to eke out a subsistence living, or simply to starve to death.

The general question that arises is:

'Can we or should we conduct our affairs in such a way as to strike a balance between encouraging safety and ensuring that the futures of those concerned are not jeopardized, particularly in the poorer countries? Alternatively, are we prepared to pay for safety standards to be imposed worldwide by paying more for commodities like uranium?' ■

Once the uranium has been extracted from the rock, the waste product (the 'tailings') contains uranium decay products, and is thus mildly radioactive. The environmental risks from mine tailings can then be greatly reduced by covering them with a few metres of earth, or by returning them underground.

Summary of Section 5.1

1   There are five basic stages in the nuclear fuel cycle: uranium mining and extraction, fuel preparation, fuel consumption in the reactor, fuel reprocessing and waste storage/disposal.

2   In principle, reprocessing, which separates the uranium, plutonium and fission products in spent fuel from each other, can be omitted, and the spent fuel treated as nuclear waste.

3   The main hazards arising from uranium extraction come from the radioactive daughter products of uranium, the most important of which is the gaseous isotope radon-222.

4   The main risk from radon during mining is to those working underground, and this risk can be reduced by adequate ventilation. The environmental risk from uranium mining waste can be minimized by burying it.

5.2 Production of nuclear fuel

As you have seen, the raw material for the production of nuclear fuel is yellowcake, which may be thought of as impure uranium trioxide, UO_3. The first step is to purify it, and this yields a free-flowing orange-yellow powder.

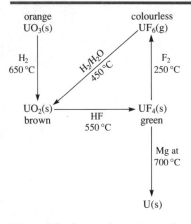

Figure 5.3 Some of the chemical processes mentioned in the text, which are important in the preparation of nuclear fuel.

Nuclear fuel is produced from purified UO_3 by using the chemical changes in Figure 5.3. The trioxide is first heated in a stream of hydrogen gas at 650 °C. This yields the chocolate-brown uranium dioxide:

$$UO_3(s) + H_2(g) \longrightarrow UO_2(s) + H_2O(g) \tag{5.1}$$

▷ There is just one type of nuclear reactor that can use the UO_2 produced at this stage as a fuel. Which one is it?

▶ You saw in Activity 3.1 that the Canadian CANDU reactor is fuelled by *natural* UO_2.

The other nuclear reactor that uses an unenriched uranium fuel is the British Magnox reactor, but in this case it is in the form of uranium metal. This is obtained by a two-stage process, the first of which involves heating UO_2 with gaseous hydrogen fluoride to produce the lime-green uranium tetrafluoride:

$$UO_2(s) + 4HF(g) \longrightarrow UF_4(s) + 2H_2O(g) \tag{5.2}$$

Heating UF_4 with magnesium metal turnings then generates uranium metal, which, because of its density, separates out cleanly at the base of the furnace:

$$UF_4(s) + 2Mg(s) \longrightarrow U(s) + 2MgF_2(s) \tag{5.3}$$

However, most of the world's reactors, including the AGR, PWR and RBMK types discussed in Chapter 3, use UO_2 enriched in ^{235}U as fuel. This enrichment is achieved by an ingenious combination of physics, chemistry and engineering.

5.2.1 Uranium enrichment

Only two methods have been used for the large-scale production of enriched uranium: these involve either gaseous diffusion or gas centrifugation. In both of them, it is essential that the uranium compound that is fed in should be a gas at, or just above, room temperature. To find such a substance was not easy, because the compounds of heavy metals like uranium are usually solids, or at least liquids, at normal temperatures. A rare exception is uranium hexafluoride, UF_6, which is gaseous above about 50 °C. It is made by heating UF_4 in fluorine gas at 250 °C:

$$UF_4(s) + F_2(g) \longrightarrow UF_6(g) \tag{5.4}$$

Enrichment using gaseous diffusion

Enrichment by **gaseous diffusion** exploits the fact that gas molecules of different mass pass through porous materials at different rates; the lighter the molecule, the faster it goes through. Thus, if the mixture of gaseous $^{238}UF_6$ and $^{235}UF_6$ obtained via Equation 5.4 is fed into one side of a porous membrane (the input side), the molecules of $^{235}UF_6$ will pass through the membrane more quickly than those of $^{238}UF_6$.

The gas mixture leaving the membrane (the output side) will therefore have a higher ^{235}U content.

The chief problem with gaseous diffusion is that after a single pass through the membrane, the enrichment is very small. For example, if we start with UF_6 made from natural uranium, it will contain 0.720% $^{235}UF_6$. When this material traverses its first membrane, the $^{235}UF_6$ content on the output side only rises to 0.723%, an increase by a factor of 0.723/0.720 or 1.004. This quantity is called the **degree of enrichment**. Because the degree of enrichment is so small, the enrichment step has to be repeated many times, and the UF_6 passes through a whole sequence of membranes or stages called an **enrichment cascade**. A gaseous diffusion cascade must contain hundreds of stages; this makes the plant very large and the process very energy intensive. Consequently, the method is being superseded by a completely different process, namely **gas centrifuge** enrichment, which we shall now describe.

Gas centrifuge enrichment

If gaseous uranium hexafluoride is placed in a cylindrical container, which is rotated about its axis at very high speeds (c. 1 000 rev s^{-1}), then the gas molecules of the heavier isotope, $^{238}UF_6$, tend to move towards the wall of the container more readily than molecules of the lighter isotope, ^{235}U (Figure 5.4). The gas nearest the wall will therefore have a higher concentration of the heavier isotope than the original mixture, and the gas nearest the axis will have a higher concentration of the lighter isotope.

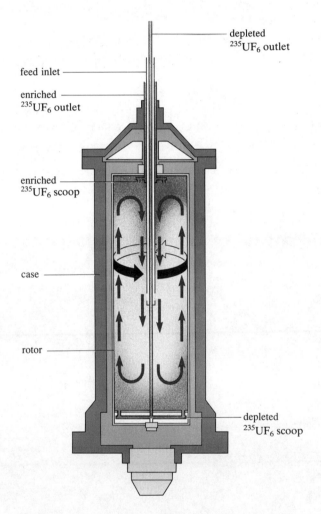

Figure 5.4 Diagram of a gas centrifuge for enrichment of ^{235}U: the arrowed cyclic gas motion (colour) down the axis and up the walls is unconnected with the centrifugal gas rotation (black arrow). It is induced separately and improves separation.

▷ Are the molecules in a gas stationary, or do they move around?

▶ The molecules in a gas move around continuously. Their speed depends on the temperature of the gas; the higher the temperature, the higher is their speed. This is called *thermal* motion.

Because of this thermal movement, there is continuous mixing of the gas molecules. As a result, although the heavier molecules *tend* to move towards the wall of the rotating vessel (rotor), this is counteracted by their general, thermal, movement. Consequently, the separation of two gases using a centrifuge is far from perfect. Nevertheless, if the gas nearest the cylinder wall is removed, it will be richer in the higher mass gas; the converse is, that the material furthest from the wall is richer in lower mass material.

As before, the efficiency of the separation process is measured by the degree of enrichment, α, which in this case, is defined as:

$$\alpha = \frac{^{235}U \text{ percentage at the outlet of the centrifuge}}{^{235}U \text{ percentage at the inlet to the centrifuge}} \qquad (5.5)$$

Here, the degree of enrichment depends on the peripheral speed of the wall of the cylinder; the greater it is, the greater is the enrichment obtained. The degree of enrichment obtained for uranium in a single gas centrifuge at lies between about 1.05 and 1.3, which is significantly larger than that for one stage of a gas diffusion plant (1.0043).

▷   If the degree of enrichment is 1.3, what is the percentage of ^{235}U produced in the uranium in the enriched output gas when UF_6 made from natural uranium is fed through a single gas centrifuge?

▶   From Equation 5.5, the fraction of ^{235}U in the output gas will be $(0.72 \times 1.3)\%$ or 0.94%.

Because the degree of enrichment in a single stage is relatively high, the number of stages needed in a gas centrifuge cascade for a given enrichment is far fewer than in a gaseous diffusion cascade. For example, in order to enrich natural uranium to a ^{235}U content of 3.0%, a centrifuge enrichment cascade need contain only six stages. However, the mass of material that can be enriched in such a cascade is small. In order to push more material through, a complete plant will have several hundred cascades running in *parallel* (Figure 5.5).

Figure 5.5   A gas centrifuge enrichment plant for uranium.

Once the UF_6 has been enriched to the required extent, it can be converted to enriched UO_2 by heating it with hydrogen and steam at 450 °C (Figure 5.3):

$$UF_6(g) + H_2(g) + 2H_2O(g) \longrightarrow UO_2(s) + 6HF(g) \qquad (5.6)$$

Enriched UO_2 made in this way, fuels most of the world's nuclear reactors. It is usually fabricated for this purpose into pellets and encapsulated in cladding. The processes involved in fuel production are summarized in Figure 5.6.

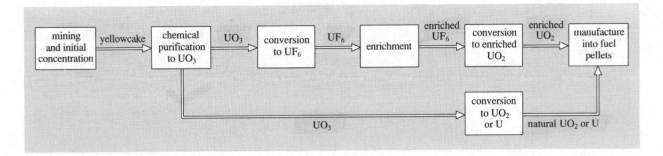

Figure 5.6 Stages in nuclear fuel production.

Although the development of gas centrifuges involves some very specialized engineering and materials, the complete systems are far smaller, and use less energy, than gaseous diffusion plants. In contrast to gaseous diffusion systems they also have the advantage that they are economical on a small scale—that is, with only a small annual output of enriched material.

The *waste product* from both the gaseous diffusion and centrifuge enrichment processes is uranium hexafluoride in which the uranium content is only 0.25 to 0.35% ^{235}U. Although a number of other techniques for uranium enrichment are under investigation worldwide, including the use of lasers, gas centrifuges currently provide the most efficient and economical method of enrichment.

5.2.2 The risks from fuel fabrication

It is generally agreed that radiation exposure of either the public or workers as a result of uranium fuel fabrication is very small. Indeed, the chemical properties of the substances involved pose a greater risk.

For example, fluorine and hydrogen fluoride, HF, are very hazardous chemicals. Fluorine, a yellow gas consisting of diatomic molecules, F_2, is the most reactive of all the elements. It reacts, often explosively, with almost all other elements. Both fluorine and hydrogen fluoride attack glass, and in contact with skin, they cause severe burns which require special treatment. In this respect, hydrofluoric acid, a solution of HF in water, is much more dangerous than more familiar acids whose burning effects cease when they are washed off the skin. Rapid and deep penetration of the tissues occurs, but the effects are delayed, and it may be several hours before intense pain develops at the site of the burn. Progressive destruction of tissue takes place and the affected part may ultimately become gangrenous (Figure 5.7). Proper medical treatment involves special injections around the burnt area. It is remarkable that the nuclear industry successfully handles large tonnages of these two ferocious chemicals each year.

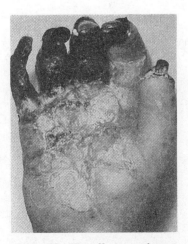

Figure 5.7 The effect on a human hand of approximately four minutes exposure to a 60% solution of hydrogen fluoride in water.

Summary of Section 5.2

1 The production of nuclear fuel from yellowcake begins with conversion to pure uranium trioxide, UO_3.

2 If enriched uranium fuel is needed, UO_3 is turned into uranium hexafluoride, UF_6, which is a gas above 50 °C.

3 Enrichment of UF_6 then occurs by gaseous diffusion or gas centrifugation. The latter is much the more efficient, having a much higher degree of enrichment at each stage.

4 The enriched UF_6 is converted into enriched UO_2 for fuel by heating it with hydrogen and steam.

5 The principal waste product from the enrichment process is uranium hexafluoride in which the uranium is depleted in ^{235}U. The radiological risks from uranium enrichment and fuel fabrication are very small, and are outweighed by the chemical risks.

Activity 5.1 *You should spend up to 20 minutes on this activity.*

Mathematics, it has been said, is a language that we turn to when our meaning can be expressed much more concisely in mathematical symbols than in words. In this activity, you will transform some wordy ideas based on Section 5.2.1 into a mathematical formula, and use that formula to calculate useful information.

At each stage in a gaseous diffusion or gas centrifuge enrichment plant, the percentage of ^{235}U in the uranium is enhanced by the degree of enrichment. Thus, if the input stream into a gas centrifuge contains 0.72% ^{235}U, the percentage of ^{235}U in the output stream (0.94%) is obtained by multiplying the input percentage by the degree of enrichment, which in this case we shall take to be 1.3—that is, at the high end of the range. In practice, an input stream passes through a linear cascade of a number of stages, the ^{235}U percentage being enhanced at each stage. So what will be the output percentage then? The following questions are designed to lead you gradually to the answer.

(a) Suppose the output stream from the gas centrifuge just discussed, containing 0.94% ^{235}U, passes into the next centrifuge in the linear cascade. What will be the output percentage from that?

(b) Suppose that the output stream from this second centrifuge is now led into a third centrifuge. What will be the output percentage from that? Compare your value with the original input of 0.72% ^{235}U. How are the two values related arithmetically?

(c) Now let us generalize using mathematics. Suppose that we call the initial percentage of ^{235}U on first entry into the cascade, P_i, and the final output percentage of ^{235}U at the end of the cascade, P_f. Suppose also that the degree of enrichment at each stage is α, and the number of stages in the cascade is n. Develop a mathematical formula that gives the final percentage P_f in terms of P_i, α and n.

(d) Suppose that $P_i = 0.720\%$, the percentage of ^{235}U in *natural* uranium. Use your formula, enrichment factors of 1.3 for gas centrifugation and 1.0043 for gaseous diffusion, together with a calculator with a y^x key, to calculate the final percentage of ^{235}U in the output stream:

 (i) after a gas centrifuge cascade of 6 stages; 3.48 %

 (ii) after a gas centrifuge cascade of 18 stages; 81.0 %

 (iii) after a gaseous diffusion cascade of 18 stages; 78 %

 (iv) after a gaseous diffusion cascade of 1 100 stages. 81 %

Comment on the relative effectiveness of the gaseous diffusion and gas centrifuge techniques.

$P_f = P_i \alpha^n$

5.3 Radioactive discharges during normal reactor operation

During routine operation of nuclear reactors, small amounts of radioisotopes are unavoidably released into the general environment. These are either fission products that have leaked from the fuel, or are radioisotopes formed by neutron interactions with the coolant. The discharges that occur during reactor operation depend on the reactor type. We shall discuss some examples from a PWR, because of the worldwide importance of these reactors.

When a reactor is operating, pinholes may develop in the cladding of the fuel pins (for example due to failure of welds in the cladding), which allows some of the gaseous fission products to escape.

▷ If fission products leak through the cladding in a PWR, where do they go to?

▶ They would enter the coolant (see Figure 3.10).

Soluble fission products leaking into the coolant can be removed by passing the cooling water through a chemical clean-up plant. Some of the gaseous fission products carried in the coolant—notably the noble gases, krypton and xenon—are discharged into the atmosphere. The main hazard in this case is from ^{85}Kr, because of its relatively long half-life (10.8 years).

As you have already seen (Section 3.2.2), a PWR has to use enriched fuel because neutrons are captured by hydrogen in the light-water coolant, generating deuterium, $^2_1$D (or $^2_1$H). If this deuterium (or that normally present in ordinary water) then captures another neutron, a new isotope of hydrogen is formed, $^3_1$H , which is called **tritium**. Although deuterium is stable, tritium has a half-life of 12.3 years, and decays by β-particle emission. Because they are low-mass gases, hydrogen, deuterium and tritium are difficult to contain (they can even penetrate some metals) and some of the tritium produced by neutron absorption in deuterium enters the environment as a gas. Tritium can also replace normal hydrogen in water molecules, in which form it might be discharged with waste cooling water.

A small proportion of water molecules contain the rare oxygen isotope, ^{17}O (abundance 0.037%). Neutron absorption by this isotope in the PWR coolant water leads to the following nuclear reaction, in which an α-particle is emitted:

$$^1_0n \ + \ ^{17}_8O \ \longrightarrow \ ^4_2He \ + \ ^{14}_6C \qquad\qquad (5.7)$$

The main product is carbon-14, which has a half-life of 5 730 years, and undergoes β-decay. After reaction with the water, it is released as gaseous CO_2 or hydrocarbons.

The most important aspect of such liquid and gaseous discharges is that they will expose the general population to a *collective dose*.

▷ What is the collective dose, and in what unit is it measured?

▶ From Section 4.3, you should recall that it is the sum of the doses (in Sv) to the group in question, and its unit is the man sievert (man Sv). If the collective dose is over a particular period (for example a year), this is also specified.

The annual collective doses arising from a proposed PWR at Wylfa, North Wales, are estimated to be, *for the whole UK population*:

liquid discharges	0.016 man Sv yr^{-1}
gaseous discharges	0.027 man Sv yr^{-1}
total	0.043 man Sv yr^{-1}

(*Wylfa-B: Environmental statement*, CEGB, 1989).

To put this in perspective, the collective dose for the operating staff (which typically number about 400) on a PWR is 1.0–5.0 man Sv yr^{-1}.

In order to interpret such collective doses, you need an idea of what the *risks* are from these exposures. Let us try to clarify this by using the risk estimates of Section 4.6. As you saw there, the low-dose health risk is cancer, and for the general population in the UK, the latest ICRP estimates suggest that there is an individual risk of 5% per sievert of developing a fatal cancer from a low instantaneous dose. Another way of looking at this is to ask what collective dose will give rise to one extra cancer death among the general population. To answer this, let us suppose that the general population consists of N people.

▷   If there is one extra cancer death among these N people, what will be the risk to the average individual of incurring this death?

▶   From the discussion in Box 4.3, it will be $1/N$ or 1 in N.

Suppose that the average individual dose corresponding to this level of risk is D. Then, according to the ICRP estimate, the risk to that average individual of developing a fatal cancer is $D \times (5/100)$. If, as we have specified here, this risk is also $1/N$,

$$\frac{5D}{100} = \frac{1}{N}$$

Therefore

$$D = \frac{20}{N} \text{ Sv}$$

▷   What is the *collective* dose for the population if each individual receives a dose, D?

▶   As there are N people, each of whom receives an average dose of $(20/N)$ Sv, the collective dose is $(N \times 20/N)$ man Sv or 20 man Sv.

Thus, the latest ICRP estimate of risk from low doses implies that there is one excess cancer death for every 20 man Sv. Although, as noted in Activity 4.2, such estimates strictly depend on how quickly the doses are received, we shall use this value of 20 man Sv per excess cancer death as a guide figure for later discussion.

▷   Use the value of 20 man Sv for the collective dose, along with an assumed collective dose of 1.5 man Sv yr^{-1} for the operators on a PWR to estimate how many radiation-induced cancer deaths would arise among the operating staff over the 30-year life of the reactor.

▶   The collective dose to the operating staff over 30 years will be (1.5 man Sv yr^{-1} × 30 yr) = 45 man Sv. If 20 man Sv produces one excess fatal cancer, then the additional number of fatal cancers among the staff as a result of 30 years of operation is: (45 man Sv/20 man Sv) = 2.25.

Whether or not you think two extra cancers is a lot depends on judgements and attitudes outside the realm of science, but it is certainly relevant to consider how many workers are at risk. As we noted earlier, there are about 400 workers at a PWR station. If two of them contract a fatal radiation-induced cancer over the 30-year lifetime of the reactor, then there will be an *average* excess risk of 2 in 400 of contracting a cancer, that is 1 in (400/2), or 1 in 200. Expressed as a percentage this is (1/200) × 100%, which is 0.5% per person.

▷ Why might considering an *average* risk be misleading to the individuals concerned?

▶ Because the radiation doses to each worker will be different. Those who absorb higher doses, for example, by working longer in the station, are more at risk than those receiving the lower doses.

Question 5.2 Use the values given in the text to calculate how many members of the UK population would die from a radiation-induced cancer due to releases from a PWR that might be built at Wylfa during the 30 years of its operating lifetime.

There are over 160 000 cancer deaths *annually* in the UK. Assuming both that the CEGB's estimates of collective doses *and* that the estimate of risk are correct, the answer to Question 5.2 shows that the public would not appear to have much to worry about from radioactive discharges if a PWR were to be built at Wylfa.

Summary of Section 5.3

1 During normal reactor operation there are releases of radioisotopes such as ^{3}H, ^{85}Kr and ^{14}C, which result in radiation exposure to the public.

2 Reactor workers are exposed to these releases, and also to radiation from the reactor and its fuel.

3 The current estimate of the collective dose leading to one excess cancer death is 20 man Sv. This implies that over the lifetime of a PWR, with a staff of 400 and a collective dose of 45 man Sv over 30 years, there will be about two excess cancer deaths among the workers. The risk to the *public*, however, is very small.

5.4 The spent fuel

Once in the reactor, the fuel lasts for up to 6 years. It is removed before *all* the fissile material has been used up, because some of the accumulating fission products absorb neutrons strongly, and so begin to impede the chain reaction. After removal, it is called **spent fuel**. How has its composition and radioactivity changed during its time in the reactor?

In the core of a PWR there is about 110 t of enriched uranium, present as the oxide UO_2. This uranium, as you saw in Section 3.2.2, is about 3% ^{235}U and 97% ^{238}U. Table 5.1 shows the changes that have taken place by the time that the fuel is removed from the reactor.

Table 5.1 The fate of the uranium isotopes in PWR fuel during their time in the reactor. Percentages by mass are given.

	Fresh fuel	Spent fuel
^{235}U content	3%	0.9%
^{238}U content	97%	95%
fission product content	—	2.5%
plutonium content	—	1.6%
total activity (110 t of fuel)	1.7×10^{12} Bq	5.8×10^{20} Bq

▷ About 2.5% of the uranium is converted to highly radioactive fission products. Which uranium isotope is most involved in this process?

▶ The fissile isotope, ^{235}U, whose percentage falls from 3% to 0.9% (Section 2.2.2).

By contrast, the fall in the ^{238}U percentage is mainly due to the neutron absorption reaction in ^{238}U. This leads to the formation of ^{239}Pu (Equation 2.12), as indicated in Figure 5.8. However, because the fuel spends a long time in the reactor, some of the ^{239}Pu may undergo one of two further nuclear reactions. Because ^{239}Pu is fissile, one such reaction is fission. The result of the other type of reaction, neutron capture, is shown in Figure 5.8; it results in the formation of ^{240}Pu.

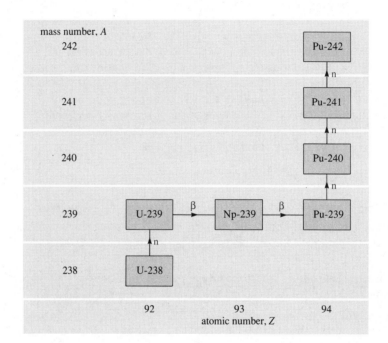

Figure 5.8 The production of plutonium isotopes from ^{238}U in nuclear fuel.

Figure 5.8 also shows that further neutron absorption reactions by plutonium nuclei will in turn generate some ^{241}Pu and ^{242}Pu. Thus, the plutonium present in spent fuel consists of a mixture of the isotopes ^{239}Pu, ^{240}Pu, ^{241}Pu and ^{242}Pu. Table 5.1 shows that this isotopic mixture comprises 1.6% of spent fuel. Uranium (element 92) and plutonium (element 94) are members of the **actinide** series of elements, with atomic numbers 89–102. Although actinides other than uranium and plutonium are produced in spent fuel, they are present in such small quantities that, as Table 5.1 implies, spent fuel may largely be considered to be uranium, plutonium and fission products.

Apart from composition, the other important difference between spent and fresh fuel is in activity. As Table 5.1 shows, the spent fuel is 100 million times more radioactive. Most of the activity is due to the fission products. Thus, the fission products account for 4.4×10^{20} Bq of the 5.8×10^{20} Bq in Table 5.1—that is, 76% of the total. The remainder is due to the plutonium isotopes and other actinides.

5.4.1 Spent fuel and nuclear waste

As you saw in Section 5.1, when spent fuel is removed from the reactor, it follows one of two routes shown in Figure 5.1. In one route, it is reprocessed in stage 4 to separate the uranium, plutonium and fission products from each other; the fission products, together with small residual quantities of actinides, then comprise the highly radioactive nuclear waste, which must be stored and ultimately disposed of in stage 5. The other route proceeds directly from stage 3 to stage 5, reprocessing is omitted, and complete fuel elements are treated as highly radioactive nuclear waste.

At this point then, we have entered the part of the nuclear fuel cycle which is concerned with the management of nuclear waste. Radioactive wastes differ in the hazard they represent. It is not just highly active wastes within the spent fuel which are involved: we can distinguish three different categories, called low-, intermediate- and high-level wastes. These classifications, which are not rigidly fixed, depend on a number of factors, including the activity per unit mass or volume, the half-lives of the radioisotopes, the way in which they decay and whether or not the decay produces so much heat that the wastes require cooling.

Low-level waste includes laboratory clothing which has become contaminated, used paper towels and liquid, gaseous and solid wastes from different parts of the fuel cycle. An important common factor that these items share is low activity and low heat production: the heat produced in them is negligible.

Intermediate-level wastes have higher activities per unit mass or volume, and hence constitute a greater radiation hazard. They do not require cooling when they are stored, but their storage has to be more elaborate than for the low-level wastes. They include fuel cladding and wastes from different stages of fuel reprocessing.

High-level waste on the other hand, produces so much heat from the decay of its radioisotopes that it requires cooling, and its safe storage requires elaborate precautions to be taken. Most of the high-level waste (97%) consists of fission products.

In the UK, the disposal of low- and intermediate-level wastes is the responsibility of NIREX (Nuclear Industrial Waste Management Executive), a government company. Its *long-term* proposals for both types of waste are that they should be buried in underground caverns or tunnels. However, the disposal of high-level waste represents the greatest challenge, as we shall see in Section 5.7.

Having discussed the types of waste that must be managed during the treatment of spent fuel, we can now discuss the two ways of dealing with it. We begin with the one that involves reprocessing.

5.5 Fuel reprocessing

Spent uranium *metal* fuel from Magnox reactors has been reprocessed at Sellafield in the UK for decades, together with fuel from military reactors (at Calder Hall and Chapel Cross), which are used to make ^{239}Pu for weapon use; we shall discuss this further in Chapter 8. However, it is now the reprocessing of UO_2 fuels which has the greatest importance, because these are the most widely used.

▷ Which reactors use uranium dioxide as fuel?

▶ Table 3.2 indicates that the AGR, PWR and RBMK reactors use UO_2 fuels.

In the UK, British Nuclear Fuels expected to start reprocessing oxide fuels from both British and foreign reactors in 1993 at the THORP (THermal Oxide Reprocessing Plant) scheme at Sellafield. In discussing reprocessing, it is important to know what form the waste products from the process take, since their safe disposal is an essential part of the nuclear fuel cycle (Figure 5.1). However, as with fuel fabrication, the *details* of the reprocessing need not concern us.

After it has been removed from the reactor, spent fuel is stored on the reactor site for a period of a year or more, to allow the decay of radioactive fission products with short half-lives. This reduces the hazard to the workers responsible for handling and transporting the waste later on.

The spent fuel is then transported from the reactor to the reprocessing plant in massive steel flasks, normally by rail (Box 5.2), where it is again stored. Both at the reactor and the reprocessing plant the storage takes place under water (in ponds), which are typically 10 m deep. The water provides cooling for the fuel, so that it does not overheat as a result of the radioactive decay of the fission products. It also provides shielding against γ-radiation.

Box 5.2   The safety of transport flasks

5.2

Many concerns have been expressed about the safety of the transport of highly active spent fuel and of the container flasks in which it is carried. One of the more spectacular tests conducted on the flasks was made by the former Central Electricity Generating Board, who put one of their 25 t spent-fuel transport flasks in front of a 140 t locomotive travelling at 160 km h^{-1} (100 mph). The flask remained intact.

John Fremlin in his book *Power Production: What are the Risks?* (Adam Hilger, 1989)

concludes that:

> *To get a genuinely dangerous radioactive cloud, the flask has to be broken open* ***and*** *the fuel taken up to 1 100 °C or so … It would make more sense to worry about even such minor hazards as chemical tankers than about (the transport of) spent nuclear fuel.* (p. 207)

You may differ from this view, and think that there *is* cause for concern. ■

As a result of minute holes in the fuel cladding, some fission products leak into the cooling-pond water, both at the reactor site and at the reprocessing plant, and the water has to be treated to remove these. This produces low-level radioactive wastes, which have to be disposed of.

The main reprocessing stages that follow the storage in the pond are summarized in Figure 5.9. The fuel pins are chopped up and dissolved in nitric acid. This releases the gaseous fission products from the fuel. Some of the gaseous fission products (notably noble gas isotopes such as ^{85}Kr) do not go into solution, and have to be collected and disposed of. Intermediate-level wastes arise from the fuel cladding and non-gaseous fission products insoluble in nitric acid.

The next stage is the separation of the fission products from the uranium and plutonium. The aqueous solution containing the fission products, which form the main *high-level* waste, is then stored again, to allow further radioactive decay to take place. The volume of fission product solution is small; about one tonne of fission products is produced annually by a typical PWR reactor with an electrical output of about 1 000 MW.

In order to allow them to be stored with the maximum safety, the aqueous solutions of high-level fission product wastes will be **vitrified** after some years; that is, they will be converted into solids by evaporation, and the solids will then be incorporated into glass. We shall be discussing the question of the treatment and disposal of high-level waste in Section 5.7.

▷   Are any special precautions needed when all these chemical separation procedures are being carried out?

▶   The workers have to be shielded from the radiation emitted by the fission products. In particular, very thick shields are needed to protect the workers from the γ-rays emitted, and all the processes are remotely controlled.

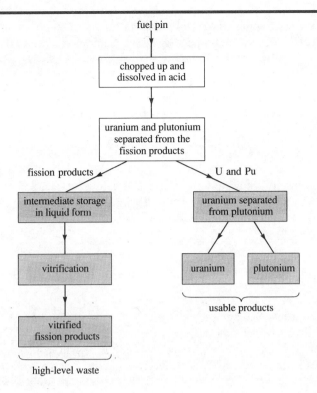

Figure 5.9 The stages in spent fuel reprocessing.

Turning to Figure 5.9 again, you will see that, after being separated from the fission products, the uranium and plutonium are themselves separated. At present, as the discussion at the beginning of this chapter implied, nearly all of this uranium and plutonium is simply stored. They could, however, be used if uranium became scarce: the plutonium and ^{235}U could be used to make reactor fuel, and the ^{238}U used as a source of fertile material for the production of ^{239}Pu in breeder reactors (Section 3.4).

At each stage of the reprocessing, liquid and solid low-level wastes are produced. In fact, although the high-level waste contains 97% of all the radioactive products, it only occupies a small fraction of the total volume of waste. Figure 5.10 shows the relative proportions of each waste category compared to the volume of the spent fuel.

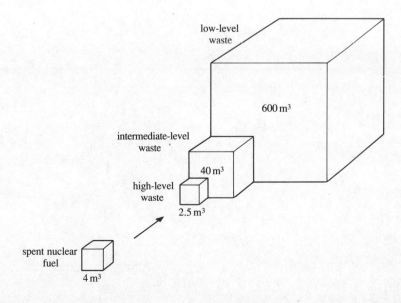

Figure 5.10 The volumes of waste produced annually from reprocessing the spent fuel from a PWR producing 1 000 MW of electricity.

▷ Calculate the total volume of waste produced from the reprocessing of 4 m³ of spent PWR fuel. How many times larger is it than the volume of the spent fuel from which it came?

▶ The total volume of the waste products is (2.5 + 40 + 600) m³ = 642.5 m³. The volume of the original spent fuel was 4 m³, so that the *total* volume of waste is a factor of (642.5 m³/4 m³) = 161 times greater than that of the spent fuel.

The total annual volume of waste from one reactor is not large: it is about equal to that of a semi-detached house. By contrast, a coal-fired power station producing 1 000 MW of electricity would use about 3 million tonnes of coal per year, and leave about 300 000 m³ of ash to be disposed of. This is some 470 times the volume of the waste produced annually by the nuclear power station; enough to fill a large housing estate in fact!

5.5.1 Radioactive waste discharges during reprocessing

Low-level wastes are produced at all stages of fuel reprocessing. Some are gaseous, and are discharged into the atmosphere. About 25 000 m³ of solid low-level wastes are produced annually at British Nuclear Fuels' Sellafield fuel reprocessing plant, and these are currently disposed of by placing them in concrete-lined trenches and covering them with soil; the site for this is at Drigg, Cumbria. The declared long-term aim is to dispose of these wastes in more secure underground depositories, which we shall describe later.

It is the disposal of *liquid* low-level wastes which has given rise to the greatest criticism in the UK. From the Sellafield site they are pumped out into the Irish Sea through two 2.5 km long pipes. The criticism has focused on several aspects, including the fact that (i) as a result of imperfections in the separation process of Figure 5.9, the discharges contain small quantities of isotopes of plutonium (which are all α-emitters), and (ii) the currents in the Irish Sea result in the widespread dispersal of the radioactivity.

As a result of these concerns, the amounts of both α- and β-emitters discharged have been reduced dramatically since the mid-1970s (Figure 5.11). Much of the radioactivity in liquid wastes is now removed by chemical treatment, and is then stored as solid waste.

Figure 5.11 The annual discharges of (a) α- and (b) β-particle emitters into the Irish Sea from the Sellafield works of British Nuclear Fuels, expressed in terms of activities. These discharges have only arisen from the reprocessing of uranium metal fuel at Sellafield; they do not include any discharges from the reprocessing of oxide fuels, which was due to begin in 1993. Note that 1 TBq = 10¹² Bq.

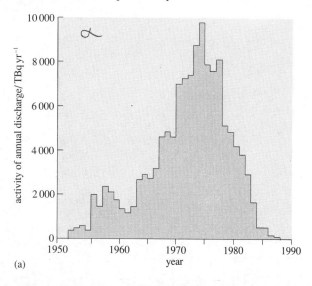

(a)

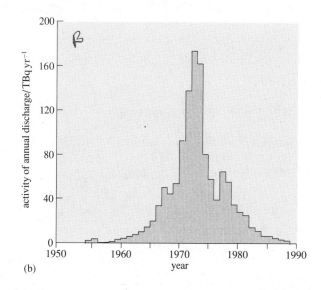

(b)

5.5.2 *Assessing the radiation risks to the population from marine discharges*

When radioisotopes are released into the sea, they eventually become incorporated into the marine food chain or in sediments. They can then enter the human food chain when seafood is eaten, or when sediments are washed ashore by currents.

In assessing the risks of such radioactive discharges, particular groups of people are identified, namely those who eat substantial quantities of the different seafoods. One group of special interest for Sellafield's discharges are people who apparently each eats 16.5 kg of *winkles* per year, in addition to 6.5 kg of crab and lobster, and 36.5 kg of fish—hardly a typical diet! Their estimated individual annual radiation dose from the radioisotopes these seafoods contain was 0.15 mSv in 1989. A similar radiation dose is experienced by a group who eat large quantities of locally grown vegetables.

▷ How does this dose from contaminated food compare with the annual radiation dose from *natural* background radiation in the UK?

▶ The average annual radiation dose from natural radiation in the UK is 1.87 mSv (Activity 4.1).

So, even given the *enormous* quantities of seafood which this particular group eats, the radiation dose they receive is less than one tenth that of the natural background radiation.

However, in addition to the radiation dose to *selected* groups, the total, *collective* dose to the whole population has to be assessed. In making this assessment, assumptions must be made about how the radioisotopes are distributed after discharge. The risk must also be calculated over the lifetimes of the isotopes that are known to have been discharged, some of which have half-lives of thousands of years. The figures we are going to give are for the collective doses to different populations *totalled* (*not* averaged) over 10 000 years following one year of discharge.

For the total marine discharges from Sellafield during 1989, for example, the estimated collective dose to the UK population was 4.5 man Sv over 10 000 years, whereas for the whole of Europe it was 27.4 man Sv. In Section 5.3 we estimated that every 20 man Sv gave rise to one excess death from cancer.

▷ Estimate how many deaths will result over 10 000 years in the UK and Europe from the 1989 marine discharges from Sellafield.

▶ For the UK the figures give (4.5 man Sv/20 man Sv) deaths = 0.23 deaths, and for Europe as a whole (27.4 man Sv/20 man Sv) deaths = 1.37 deaths.

Only one to two extra deaths over 10 000 years in the UK and Europe is a very low figure indeed, so low that it seems incredible. In fact that was our reaction when we first calculated it, so we went back to the people who had produced the data to check that we had interpreted them correctly. The data are based on only one year's discharge: if we consider the effect of 100 years of discharges at the 1989 rate, then the collective doses and numbers of deaths will be multiplied by 100. Then, for example, there would be, for the UK, about 23 deaths in 10 000 years as a result of the marine discharges from Sellafield, if it operated at 1989 discharge levels for 100 years.

The figures for the effects of gaseous discharges are similar, so that the total extra deaths over 10 000 years would be about 50. This is *still* a very small number indeed. *If it is correct*, then you may feel that it is too small to justify taking any more precautions to reduce the radioactive emissions even further; one of the considerations that you might take into account is that the money that would need to be spent in saving 50 lives over 10 000 years would, if invested in other life-saving measures in other industries, save many more lives.

But *are* these estimates of the risks correct? And if they are *not*, how wrong are they? Let us go back and look at the basis for them. The factors on which they depend include:

(i) knowing the type and quantity of discharge;

(ii) modelling the subsequent movement of the radioisotopes and their mode of entry into the human food chain;

(iii) identifying critical groups who are particularly at risk, and

(iv) knowing what the risks of exposure and uptake are.

The first three factors are specific to the site, whereas the fourth represents fundamental information for the effects of radiation in general.

Factors (ii) to (iv) are the responsiblity of the Ministry of Agriculture, Fisheries and Food (MAFF) as well as of British Nuclear Fuels Ltd (BNFL) themselves. The effort required to undertake these assessments is substantial, and groups who are sceptical of the results, for example Friends of the Earth, limit themselves to investigating specific aspects of the predictions (for example the radiation levels in the silt of estuaries near the site), and to seeing if their data are consistent with the published values. They are also concerned with investigating whether there are any unidentified critical groups.

Such investigations have shown that there are gaps and inconsistencies in the officially reported data. For example, a survey undertaken by Friends of the Earth (FoE) in 1988–9 (*Unacceptable levels: a report by FoE's Radiation Monitoring Unit of the Sellafield contamination of the River Esk, Cumbria*, 1989) showed that there was a 6 km stretch of the River Esk where there was radioactive contamination of the river and of the adjoining fields. The contamination arose because the river is tidal. The survey suggested that if a person spent 34 hours per week (nearly a full working week) *every week of the year* in the vicinity of the river bank, he or she would receive the maximum permitted radiation dose. For releases from a single site such as Sellafield, this is *currently* 0.5 mSv yr^{-1}.

▷ If they did this, what would be their annual risk of contracting a fatal cancer?

▶ The ICRP risk estimate of contracting a fatal cancer from an instantaneous low-level radiation dose is 5% Sv^{-1}. So the risk from 0.5 mSv is

$$(0.5 \times 10^{-3})\,\text{Sv} \times \left(\frac{5}{100}\right)\text{Sv}^{-1} = 2.5 \times 10^{-5}$$

$$= \frac{2.5}{10^5} = \frac{1}{40\,000}$$

So the chance of developing a fatal cancer as a result of one year spending 34 hours every week of the year on the banks of the River Esk are 1 in 40 000; over a period of 10 years the chance would be 1 in 4 000.

There are bound to be inaccuracies and inconsistencies in anything as complicated as predicting the health effects of industrial discharges, of which the radioactive discharges from Sellafield are an example. Two important questions have to be asked:

(a) Are the *overall* conclusions about the risks involved correct?

(b) Are these risks justified by the benefit to society from the activity?

The first question is one that should, in principle at least, lend itself to being answered objectively. However, any answer to the second one is essentially subjective;

for this reason this question is the most contentious. *Neither* are questions to which there are simple answers, as you will see when we pick up these issues again in Chapter 6.

Summary of Sections 5.4 and 5.5

1 Spent PWR fuel consists of about 95% ^{238}U, 0.9% ^{235}U, 2.5% fission products and 1.6% plutonium. The spent fuel is about 100 million times more radioactive than fresh fuel; this is mainly due to the fission products.

2 The handling of spent fuel involves the management of nuclear waste, which, for the purposes of disposal, is divided into three categories: low, intermediate and high level. High-level waste requires cooling facilities.

3 Spent uranium metal fuel has been reprocessed at Sellafield for many years. From 1993 oxide fuels from British and foreign reactors began to be reprocessed in a new plant, known as THORP.

4 One year's operation of a PWR produces about $4\,m^3$ of spent fuel, which, on reprocessing, yields $2.5\,m^3$ of high-level waste, $40\,m^3$ of intermediate-level waste and $600\,m^3$ of low-level waste.

5 Current marine discharges at Sellafield are much lower than in the 1970s, and the estimated risks to the general public are small. However, there are some indications that neither the modelling nor the monitoring may be detailed enough to allow an accurate evaluation of the risks to selected groups in the local population.

5.6 Strategies for dealing with high-level nuclear waste

We have now arrived at the fifth and final stage of the nuclear fuel cycle of Figure 5.1. This is waste storage or disposal. The distinction is important. **Storage** means that the wastes are kept in a safe but accessible form, perhaps for up to 100 years, with the clear intention of then subjecting them to subsequent treatment. **Disposal** means that the waste is put into a place where it can be safely left, and from which there will be no need to recover it except, perhaps, in a serious emergency. A number of disposal methods have been considered, but the current preference is for burial in tunnels in deep geological formations, and that is the option we consider here.

Notice that, in Figure 5.1, the final stage is arrived at by one of two routes: the nuclear waste that arrives at the final stage is spent fuel, which either has or has not been reprocessed. In either case, it can then undergo either storage or disposal, so there are four possibilities in all. The two basic strategies for dealing with nuclear waste involve combinations of these four possibilities, as shown in Figure 5.12.

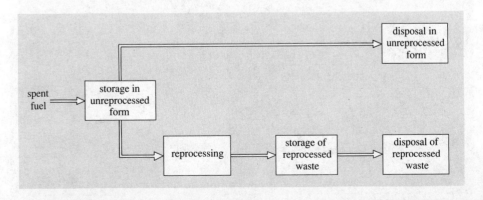

Figure 5.12 Two alternative strategies for dealing with nuclear waste.

At the present time, all spent fuel is first stored, whether it is to be reprocessed or not. Storing wastes has the advantage of allowing some of the shorter-lived radioisotopes to decay away so that the risks of later handling are reduced. It also provides time to make more considered decisions about the subsequent treatment. Initially, as you saw in Section 5.5, this storage is in ponds at the reactor site, and, if it is to be reprocessed, at the reprocessing plant as well. In the UK, nearly all nuclear waste then currently follows the lower route in Figure 5.12: the spent fuel is reprocessed, thus removing the uranium and plutonium, and the high-level waste, which now consists of the fission products, is stored again.

▷ Immediately after reprocessing, what form are the fission products in, and what form do they eventually assume?

▶ Section 5.5 and Figure 5.9 showed that the high-level fission product waste is an aqueous solution initially, and that it is eventually vitrified—turned into a glass.

At Sellafield, the initial aqueous solution is stored in double-walled, water-cooled stainless steel tanks enclosed in concrete (Figure 5.13). After some years, this solution

Figure 5.13 A storage tank for liquid high-level waste in the process of construction. Note the piping, which mainly carries cooling water.

is evaporated, and the dissolved solids heated with oxides of lithium, sodium, boron and silicon to give a borosilicate glass. The glass is then sealed in stainless steel containers and stored in air-cooled vaults near the surface.

Although there are advantages in storing waste, it is probably not practicable to maintain the storage buildings for hundreds of years. Consequently, high-level waste will probably eventually undergo disposal, with no intention of retrieval. Indeed, the current intention in the UK is for *all* nuclear wastes (low, intermediate and high level) to be eventually enclosed in special containers and buried deep underground in tunnels or caverns known as **depositories** (or **repositories**). The external radiation dose to people on the Earth's surface will then be *much less* than that from normal background radiation because of the shielding that would be provided by the rock strata above the depository.

This disposal policy will be exercised whether the waste has been reprocessed or not, so it terminates both the upper and lower strategies of Figure 5.12: in the upper one, spent fuel is first stored and then disposed of in this way; in the lower one, disposal is the terminal fate of the vitrified fission products obtained from reprocessing. Waste disposal is therefore an issue that ultimately cannot be avoided. Let us now look at some of the challenging problems to which it gives rise.

5.7 High-level waste disposal

In this section we shall discuss two key problems of high-level waste disposal. The first is the time-scale over which the waste must be managed; the second is the problem of ensuring the immobilization of the waste when it is disposed of deep underground.

5.7.1 Time-scales and heat production

How quickly *do* the radioactive wastes decay away? The picture is complex, both because of the range of different radioisotopes initially present in the waste, and because many of them give rise to radioactive decay chains, so that new radioisotopes are continually being formed.

The way in which the activity from spent fuel decays with time is shown in Figure 5.14, for both *total* activity and for some of the biologically important radioisotopes that have been discussed earlier. Note that on both axes the scales are logarithmic. There are wide differences in the rates at which the different radioisotopes decay, that is, the steepness of the different curves. In particular, the plutonium isotopes decay much more slowly than most of the other isotopes.

▷   Roughly what fraction of the total activity is due to the plutonium isotopes (i) initially, and (ii) after 1 000 years?

▶   At zero time, the activity from the plutonium isotopes is about 2×10^{15} Bq out of a total activity of 10^{20} Bq. Thus, plutonium forms a fraction of $(2 \times 10^{15}/10^{20})$ of the total activity—that is, 0.000 02, or 0.002%. After 1 000 years the plutonium activity is roughly 4×10^{12} Bq out of a total of 4×10^{13} Bq, so it forms a fraction of $(4 \times 10^{12}/4 \times 10^{13})$ which is 0.1, or 10%.

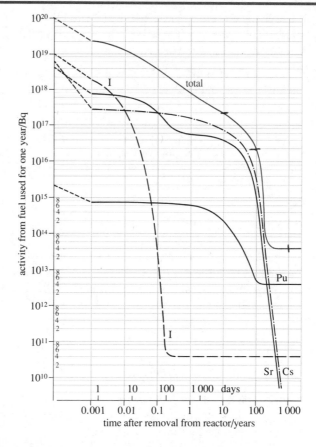

Figure 5.14 The change in activity with time of some important elements (caesium, strontium, iodine and plutonium) in the high-activity waste produced annually by a 1 000 MW(electrical) PWR after removing the spent fuel from the reactor. For reference, the activity at $t = 0$ (removal from the reactor) is indicated by the intersection of the dashed lines with the activity axis. Zero cannot, strictly speaking, appear on a logarithmic scale.

Question 5.3 From Figure 5.14, what fraction of the initial total activity at zero time will remain after 10, 100 and 1 000 years?

You can see from the answer to Question 5.3 that even storing nuclear wastes for ten years gives a significant reduction in activity. After the very rapid fall between 100 and 1 000 years, the total activity diminishes only very slowly: the curve is almost level. Because some of the radioisotopes decay so slowly, any method of storage or disposal has to keep the wastes out of the environment for at least hundreds of years. As you know, when spent fuel is removed from the reactor, heat will continue to be produced by the fission products. For this reason, the nuclear waste containing the fission products has to be cooled for tens of years, until most of the shorter half-life isotopes have decayed.

5.7.2 Water and underground waste disposal

The currently preferred policy on waste disposal is for *all* the wastes to be eventually buried in tunnels or caverns deep underground after enclosing them in special containers. Assuming that the radioactive waste *stayed* underground, the problem of disposing of radioactive waste would appear to have been solved. So, can we ensure that this happens? The possibility of earthquakes, or of geological changes generally, which could bring the waste back to the surface, are probably the first things one thinks of. However, although these capture the imagination, they are not the main problem, which is simply *water*.

The threat posed by water is twofold. Firstly, if it is able to penetrate the depository, water can corrode any containers around the waste and then dissolve the radioactive wastes inside. Secondly, the movement of water underground can *transport* any dissolved radioactive waste from its burial site to the surface. Once at the surface, the wastes can be incorporated into the food chain or dispersed into the environment.

To explore such problems, we must consider how water is distributed, and moves, around the world. Both are shown in Figure 5.15. The majority of the world's water (about 94%) is stored in the oceans, the next largest store being the groundwater (4%). You will see from Figure 5.15 that the **hydrological cycle** consists of transport of water by evaporation into the atmosphere, followed by precipitation as rain or snow, with a flow back into the oceans through both surface and groundwater.

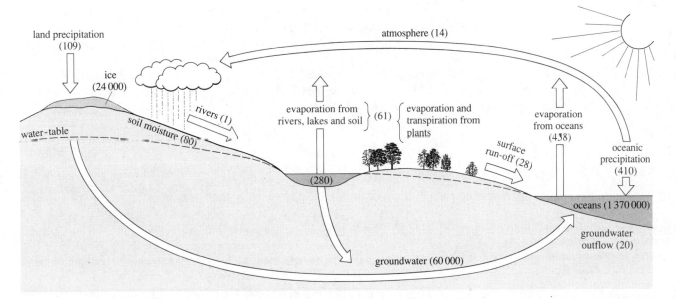

It is the transport of water into and out of the groundwater which provides the mechanism whereby radioisotopes from wastes stored underground could be returned to the environment. However, at any stage in the cycle where evaporation occurs, anything dissolved in the water is left behind. In particular, radioisotopes transported into the oceans would accumulate there; they would not evaporate and re-enter the water cycle.

Figure 5.15 The global water cycle: the hydrological cycle, showing the distribution of the world's water. All volumes (shown in parentheses) are in thousand cubic kilometres. 'Lakes' includes freshwater and saline lakes. The volumes shown in colour represent the amounts of water cycled annually, as opposed to stored (black).

Question 5.4 From the data given in Figure 5.15, what percentage of the groundwater flows into the oceans annually? What does this suggest about the rate of groundwater flow?   ○ 03%   very slow

The answer to Question 5.4 suggests that if groundwater penetrates an underground radioactive waste depository, it may only dissolve and carry away the radioisotopes very slowly. However, as you have already seen, it is only after about 1 000 years that the radioactivity from spent fuel has decayed substantially, so that ensuring the integrity of underground waste depositories for periods of thousands of years is important.

You should not get the idea that after 1 000 years the waste is safe, or that *any* release of activity before this time would be very dangerous: Figure 5.14 showed that the activity is decreasing continuously with time. The longer the radioactivity is kept out of the environment, the lower the associated risks will be.

With this background, let us look at the factors affecting the choice of a suitable underground site for disposing of radioactive waste.

5.7.3 Choosing the right rocks

Many of the rocks that make up the Earth's crust contain **voids**, which can hold water. These voids can take various forms. In sandstones, for example, they consist of small interconnected pores; in granites there may be fissures or fractures, which may or may not be interconnected. Below a certain level, the rock voids are all filled with water. This level is called the **water-table**, and the rocks below it are said to be **saturated**. By using the voids as a pathway, water can flow through the saturated rocks. When selecting a site for a waste depository, rocks in which the velocity of the water flow is small are sought. Let us now consider some of the factors that influence the flow rate.

Water only flows between two points if there is a pressure difference between them. How does a pressure difference arise in the groundwater? One way is shown in Figure 5.15, where you can see that the water-table is not level. As a result of differences in height of the landscape, and of the fact that water generally only flows slowly through the underlying rock, there are gradients in the water-table, which create pressure differences. Pressure differences are usually measured in terms of the difference in the level of water between two points. Thus, if the height of the water-table at two points separated by a distance l is H_1 and H_2 (Figure 5.16), then the **hydraulic gradient**, I, which is a measure of the pressure difference, is given by:

$$I = \frac{(H_2 - H_1)}{l} \tag{5.8}$$

Figure 5.16 The hydraulic gradient and groundwater flow.

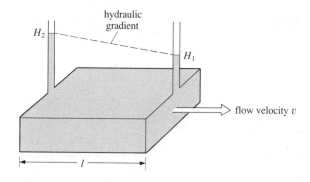

▷ What do you think the unit of hydraulic gradient is?

▶ As both the water-table height, H, and distance, l, are measured in metres, hydraulic gradient is a dimensionless quantity; that is, it is a pure number.

The velocity of passage of water through a given rock is proportional to the hydraulic gradient. We can therefore write the equation:

$$v = KI \tag{5.9}$$

where v is the velocity and K is a constant known as the **hydraulic conductivity**. Both v and K have units of velocity. The hydraulic conductivity is a characteristic of the rock: if the rock contains large well-connected pores or voids (Figure 5.17a), or extensive linked fractures (Figure 5.17b), water will flow easily through the rock, the hydraulic conductivity will be large, and flow velocities will tend to be high.

▷ Why is the hydraulic conductivity of the rock in Figure 5.17c small?

▶ The pores are not interconnected, so the water cannot flow easily through the rock. Of the three rock types shown in Figure 5.17, it is the one in part (c) which might be useful as a waste depository.

(a)

(b)

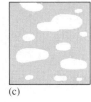

(c)

Figure 5.17 Schematic representations of three different kinds of void structure within rock. The white areas in the diagrams represent pores or voids.

Let us try to get a *feel* for the velocity with which water flows underground in the types of rock considered suitable for waste disposal: is it typically measured in kilometres per hour, metres per year or millimetres per million years?

μ ι

▷   Examples of values for hydraulic conductivity and hydraulic gradient for a clay are $8 \times 10^{-8}\,\mathrm{m\,s^{-1}}$ and 0.2, respectively. What is the velocity with which water flows through the clay?

▶   Equation 5.9 gives the velocity as *KI*. Substituting the values given, the velocity is $(8 \times 10^{-8} \times 0.2)\,\mathrm{m\,s^{-1}} = 1.6 \times 10^{-8}\,\mathrm{m\,s^{-1}}$. However, $0.000\,000\,016\,\mathrm{m\,s^{-1}}$ is fairly slow, and is difficult to imagine! Let us therefore convert it into $\mathrm{m\,yr^{-1}}$.

One year is $(365 \times 24 \times 60 \times 60)$ seconds, which is $3.15 \times 10^{7}\,\mathrm{s}$. Changing the units,

$$1.6 \times 10^{-8}\,\mathrm{m\,s^{-1}} = (1.6 \times 10^{-8}\,\mathrm{m\,s^{-1}}) \times (3.15 \times 10^{7}\,\mathrm{s\,yr^{-1}}) = 0.50\,\mathrm{m\,yr^{-1}}$$

So, the flow rate in this example is about half a metre per year. Although, as we shall see, a wide range of water flow rates is encountered in different types of rock, using the unit *metres per year* is very convenient, and gives you the right feeling for the orders of magnitude involved in the rock types that might be appropriate for nuclear waste disposal. However, in aquifers (rocks from which groundwater is extracted) flow rates may be at least hundreds of metres per year.

Although the rate of flow of groundwater through a rock type is an important factor in deciding on the suitability of a site for an underground depository, it is not the *only* one. *Chemical* interactions between the rock and the groundwater may also be important. There may, for example, be chemical mechanisms that prevent, or hinder, the flow of dissolved radioisotopes in the rock strata. One of the best examples of this occurs in clay, where positively charged radioactive ions from the water may replace non-radioactive ones at negatively charged sites in the clay. This process is called **ion exchange** and is shown in Figure 5.18a. Such a process tends to immobilize the radioactive ions. Radioisotopes in solution may also become incorporated into the matrix of the rock—a process known as **mineralization**—and so are retained within it (Figure 5.18b). This too can slow down or stop the transport of radioisotopes in water. Because such effects are chemical, they will be different for different elements.

Figure 5.18   Two mechanisms that can slow down, or stop, the transport of radioactive species. (a) Ion exchange: in this process a positively charged radioisotope ion dissolved in groundwater replaces another positive ion in a clay particle. (b) Mineralization: an atom from a radioactive species in solution occupies a vacant site within the molecular structure of the rock.

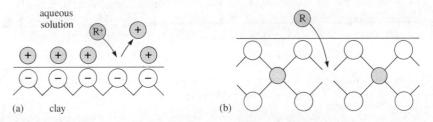

In addition to the importance of the rock *type* in determining its suitability for radioactive waste disposal, its location and general geological environment is equally important, because this will determine the overall flow pattern for the groundwater and the hydraulic pressures that can be built up. The flow pathway from the waste depository to the Earth's surface should be as long as possible, in order to maximize the time taken for any dissolved radioactive species to reach the surface environment.

▷   Why does maximizing the time taken by radioisotopes to reach the surface matter?

▶   The radioisotope will decay while it is moving to the surface, thus reducing the eventual radioactive hazard.

Rock types that have been suggested as suitable for radioactive waste disposal sites are hard igneous or metamorphic rocks, such as granite, and impervious sedimentary rocks, such as clay. However, even within these two rock types there is a wide range of values for the parameters that determine water flow—the hydraulic conductivity and the hydraulic gradient. Table 5.2 shows this clearly for four sites assumed to be more than 100 m deep.

Table 5.2 Sample hydraulic parameters for two types of rock in environments that might be considered for a radioactive waste depository at a depth greater than 100 m.

Rock type		Hydraulic conductivity/ $m\,s^{-1}$	Hydraulic gradient	Groundwater velocity/$m\,yr^{-1}$
clay	(i)	2×10^{-8}	0.2	0.13
	(ii)	3×10^{-12}	0.05	4.7×10^{-6}
granite	(i)	10^{-6}	0.1	3.2
	(ii)	10^{-8}	0.001	3.2×10^{-4}

As you can see from Table 5.2, groundwater velocities in some situations are very low indeed. In one of the two granites, the velocity is over $1\,m\,yr^{-1}$, but in the other granite and in one of the clays it is less than $1\,mm\,yr^{-1}$. With flow rates as low as this, you can see why it takes hundreds, thousands and sometimes millions of years for deep groundwater to reach the surface; you can also see why only a tiny percentage of the groundwater (0.03%) flows into the oceans annually (answer to Question 5.4).

However, Table 5.2 also shows the diversity of flow conditions within a particular rock type. Each potential site therefore has to be very carefully studied. In particular, the distribution of fissures within the rock may be unpredictable and uneven, giving rise to disturbing fluctuations in flow velocities. Knowledge of fissure size and distribution is therefore an important part of any proper site evaluation, but a great deal of drilling may be needed to achieve that information.

As water raises so many problems for storage/disposal sites, are there any rock types in which water does not appear to permeate? There are, and they are called **evaporites**—that is, rock formations made up of soluble salts (for example NaCl), which were formed by the evaporation of seawater from ancient seas. Because the salts are very soluble, the existence of these underground deposits appears to show that water has not permeated the deposits since they were formed.

The criteria for selecting a site for a radioactive waste depository are therefore:

(a) Geological stability: the site should not be in an area prone to earthquakes or volcanic activity, or where there could be changes in the groundwater flow.

(b) The groundwater flow should be slow (or non-existent), and water pathways to the surface should be as long as possible.

(c) It is an advantage if the rocks through which water passes can chemically impede, or prevent, the transport of radioisotopes from the site.

(d) In addition, the rock strata in the vicinity of the waste depository should be thermally stable; that is, their properties should not change when heated. This is because the waste will generate heat from fission-product decay.

(e) Finally, locating the depository in rock strata where the direction of groundwater flow is out to sea can ensure that any radioisotopes that do escape benefit from the dilution and dispersion that the sea can provide.

Some of the areas that are thought to fulfil these criteria in the UK are shown in Figure 5.21 (p. 112).

Question 5.5 A waste depository buried deep in the first of the two clays in Table 5.2 is breached by groundwater, which then follows a 1 300 m pathway to the surface. How long will it be before radioactive waste reaches the surface?

5.7.4 The design of a high-level waste depository

As noted in Section 5.6, a special site in which nuclear waste is disposed of is called a depository or a repository. Initially, these two terms distinguished sites with non-retrievable waste (depository) from those where retrieval was possible (repository). With time, however, this distinction has become blurred.

It is accepted that waste deposited underground cannot be isolated from the environment for all time. However, as you saw in the answer to Question 5.3, after a thousand years the majority of the shorter-lived radioactive species have decayed away, and the amount of radioactivity has fallen by a factor of over a million. So the problem is to isolate the wastes until the hazard that any releases pose is negligibly small. To achieve this, the design of the waste depository includes a number of barriers, designed to delay both the release and the transport of the waste.

A high-level waste depository would consist of a series of underground shafts, caverns or tunnels (Figure 5.19). A minimum depth of 300 m is invariably specified, and in practice, the high-level radioactive waste depositories proposed so far range in depth from between 300 m and 1 000 m.

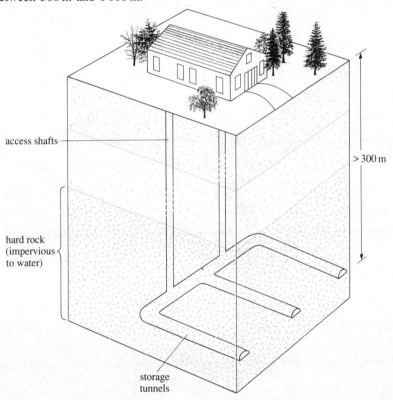

access shafts

hard rock (impervious to water)

storage tunnels

> 300 m

Figure 5.19 Digrammatic section through a proposed underground high-level waste depository.

The choice of a suitable geological site for waste disposal (Section 5.7.3) is one factor in determining safety. The second is the form in which the waste is buried, and the precautions taken within the depository itself to inhibit transport of the radioisotopes.

As you saw in Section 5.6, at present, waste in the UK is eventually vitrified by incorporating it into a borosilicate glass matrix. This glass is highly resistant to corrosion, and the radioisotopes bound within it can only be extracted very slowly by any surrounding water. Thus, vitrification of the waste provides the first barrier for preventing the radioisotopes from entering the environment.

In order to delay groundwater from reaching the vitrified waste, cylinders of it would be encapsulated in a thick metal container—for example, 250 mm of stainless steel. This provides a second barrier for preventing the escape of radioisotopes. These containers are placed in tunnels or shafts, which are then *backfilled* with a material, such as certain clays, whose chemistry inhibits the transport of the radioisotopes (Figure 5.18) and which is impervious to water flow; this is the third barrier.

If the tunnels or shafts are lined with concrete, this will provide the next barrier, together with the rock in the immediate vicinity of the tunnel walls. These are all called the *near-field* barriers (Figure 5.20). Finally, the last barriers—called the *far-field* barriers—are the rock strata between the near field and the surface, which determine the water flow into and out of the depository site, and the degree to which transport of dissolved radioisotopes is delayed, and thus are able to decay further before they reach the surface environment.

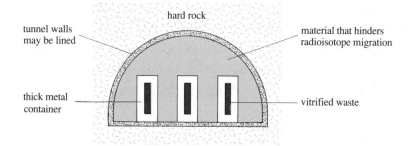

Figure 5.20   Cross-section through a tunnel for radioactive waste disposal, showing near-field barriers.

5.7.5 High-level waste disposal in the UK

Because of political opposition, there are currently no proposals for a *high-level* waste depository in the UK, although this question received considerable attention in the 1970s. In particular, a number of potential sites were identified (Figure 5.21), and

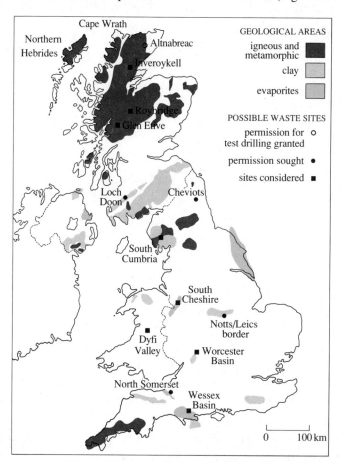

Figure 5.21   Geological sites in the UK considered as potentially suitable for high-level waste disposal in the late 1970s.

applications were considered or made for test borings to obtain detailed data on the rock strata and groundwater flows. As you might expect from Section 5.7.3, the rock types that were to have been investigated included unfissured granites (for example in Cornwall, Cumbria and Scotland), clays (in the Midlands) and evaporites (for example at Teeside).

However, although some test drilling was undertaken, the opposition to planning applications to undertake the test drilling was so strong that the government abandoned the high-level waste test programme in the early 1980s.

5.7.6 The Oklo phenomenon — a natural disposal site

So far, we have been discussing nuclear reactors as if they did not exist before man made them. This is not true! There were nuclear reactors in the world 1 800 million years ago, but these were *natural* ones. To see how this was possible, we can use the argument of Question 2.6, but on a different time-scale. In Question 2.6 we saw that, because of the difference in their half-lives, the fraction of ^{235}U which had decayed in the 4 500 million years since the Earth was formed was 0.984, whereas for ^{238}U it was only 0.5. If we were to repeat the calculation, this time looking at the fraction that had decayed in the last 1 800 million years, we would find that for ^{235}U it was 0.830 (83.0%) and for ^{238}U it was 0.244 (24.4%). Using these figures to work out the percentage of ^{235}U in 'natural' uranium 1 800 million years ago, shows that it was about 3%.

▷　What is the percentage of ^{235}U in the enriched uranium used for PWR reactors?

▶　The fuel enrichment for PWRs is 3% (Table 3.2).

You can see from this that, 1 800 million years ago, natural uranium had the same ^{235}U concentration as the fuel used in PWRs. In the rich uranium deposits at Oklo, in Gabon, West Africa, the percentage of uranium in rocks rises to 50% in some places. The fact that the rocks were saturated with water, which would have acted as a moderator, was enough to make natural thermal reactors critical in several places. The chain reaction could have been initiated by neutrons produced from the spontaneous fission of uranium-235, for example. These reactors ran for several thousand years, during which time they generated an estimated 100×10^9 kWh of energy, that is, the same as a modern nuclear reactor running for about 5 years.

How was this discovered? During a routine French laboratory testing of uranium hexafluoride derived from Oklo uranium ore, it was found that some samples had ^{235}U concentrations less than the normal 0.72% in natural uranium. The only way in which this could be accounted for was to suppose that the ^{235}U had been used up in a fission chain reaction. Subsequent investigation of the mines from which the samples came confirmed this hypothesis, in particular, by analysis of the fission products present in the ore.

What this suggests is that *the Oklo uranium deposits are the world's first high-level waste disposal site*. We thus have the opportunity to study the migration of the fission products and actinides over very long time-scales indeed, and at a site that has not been carefully chosen to have low water flows.

What has been found is that, not surprisingly, most of the noble gases (for example krypton and xenon) have been lost. However, although alkali metals such as caesium have migrated substantial distances, only a small fraction of the lanthanide elements, such as lanthanum and samarium, have moved, and then not more than a metre from their production site. The least mobile group of radioisotopes were the actinides, including plutonium, for which migration was estimated to be of the order of *millimetres* or less. Overall, only a few per cent of the fission products migrated from the source of the nuclear reaction, with a spread of several metres. Over 1.8×10^9 years, this is equivalent to a flow rate lower than the lowest groundwater velocities in Table 5.2.

Fascinating though it is, the Oklo phenomenon only gives migration data relevant to that site. Nevertheless, it provides evidence that, under some circumstances, high-level waste does not migrate significantly over time-scales *far* greater than those of interest for the disposal of man-made radioactive waste. It certainly provides a test bed for models of radioisotope transport.

Summary of Sections 5.6 and 5.7

1   There are two basic strategies for dealing with nuclear waste: firstly, storage of spent fuel followed by later disposal; secondly, storage and then reprocessing of spent fuel, followed by storage of the highly active fission-product waste and its subsequent disposal.

2   The currently favoured method of disposal is burial in geological formations deep underground.

3   High-level wastes should be allowed to decay for hundreds or thousands of years in a secure location before they can be allowed to enter the environment. For the initial decades, they must be cooled because of the heat produced by fission-product decay.

4   Waste disposed of deep underground might be mobilized by groundwater, which can move through interconnected voids in the rock.

5   The principal criterion for a high-level waste depository site is that it has a low hydraulic gradient; rocks such as clays or granites, which often have low hydraulic conductivity, fulfil this criterion. Groundwater flow velocities in these rock types are low. Evaporites, the very existence of which (since they are water soluble) demonstrates that they have not been penetrated by water, may also be suitable.

6   It may be difficult to establish the suitability of a site because the distribution of fissures, which are partly responsible for groundwater flow, may be uneven and unpredictable.

7   Proposals for high-level waste disposal currently centre on the placement of vitrified waste in depositories at depths of between 300 m and 1 000 m.

8   In such depositories there would be multiple barriers to prevent the radioisotopes from entering the environment, including encapsulation of the vitrified waste in metal containers and the use of a surrounding material that restricts transport of radioisotopes.

9   There are no firm proposals at present to build such a depository in the UK, following opposition to test programmes in the 1970s.

10 Nearly two thousand million years ago, there were natural reactors at Oklo, in Gabon. Study of the distribution of fission products and the isotopes in their decay chains has shown that very few of them have migrated far from their site of formation, particularly the plutonium isotopes.

11 Since the migration of radioisotopes is site specific, it is not possible to draw any *general* conclusions from the Oklo phenomenon.

5.8 The opposition to nuclear waste disposal

The problem of dealing with nuclear waste, like other topics in *Science Matters*, demonstrates the difficulties that scientists have in making forward projections for environmental systems. The entanglement of scientific, ethical and political questions which it engenders is also typical.

The political element intrudes most strongly when moves are made to select a site for a nuclear waste depository. There is always some opposition to any new and large industrial development, but it is often tempered by an awareness of the increased wealth and employment that the development will bring to the region concerned. But nuclear waste disposal sites, in common with waste disposal sites generally, do not even have this advantage. They offer only short-term employment while the depository is being built, and generally, people do not like living near a waste dump. Conflicts of this sort, in which local opinion must be weighed against the balance of the risks and benefits of an activity for society as a whole, are particularly taxing ones.

The effectiveness of local protest would be much diminished if the choice of site were backed by overwhelming scientific arguments. But that cannot be so. The main scientific issues are (i) whether the models used to predict the release of radioisotopes from a waste depository are detailed enough, and (ii) whether the data they require are known or can be obtained. Note that these two issues are separate; both the models *and* the data need to be reliable before the predictions can be depended on. Both aspects contain uncertainties, so their use to make geological predictions, particularly on long time-scales can always be criticized. Such arguments have been set out by Philip Richardson (*Exposing the Faults: the Geological Case against the Plans by UK NIREX to Dispose of Radioactive Waste*, Greenpeace and Friends of the Earth, 1989). He constantly emphasizes the simplifications and assumptions that underpin the predictions, raising, among other things, the question of whether the depository will change the geology of the near field, and the effects of glaciation and climate change. The conclusion of the report is that:

> *It would appear from the scale of the uncertainties highlighted in this report that the deep disposal option is not proven to be a safe and reliable method of immobilizing and isolating nuclear waste from the environment.* (p. 17)

At the very least, such uncertainties can be used to strengthen any protest campaign, and to raise the question of alternative sites. The result is that the problem of assessing potential sites is made extremely difficult, and the selection process moves away from being based only on general scientific issues to becoming a political decision. These problems are highlighted by A. G. Milnes (*Geology and Radwaste*, Academic Press, 1985), who says:

> *The Earth scientists involved often find themselves as pawns in an economic–political game, to be used or discarded as the need arises with little regard for the scientific merits of the case. Most geologists are convinced that careful and methodical study would reveal sites which would satisfy accepted safety standards, but many are uneasy about the political climate in which studies are presently being carried out.* (p. 296)

However, others see the issue completely differently. For example, in the preface to their book *The International Politics of Nuclear Waste* (Macmillan, 1991), Andrew Blowers, David Lowry and Barry Solomon say:

> *Sophisticated geological analysis, risk assessment or modelling of repository behaviour must rest on heroic assumptions and are no substitute for empirical*

knowledge. Scientific predictions for periods of 10 000 years or more lie in the realm of fantasy, not rationality. In conditions of uncertainty it must be concluded that there is no technical solution to the problem of radioactive waste. (p. xvii)

We thus have strikingly different views on whether it is possible to dispose of radioactive wastes safely.

Nuclear power is not necessarily unique in requiring its risks to be assessed over very long periods of time. If global warming as a consequence of CO_2 emissions proves to be a real phenomenon, then very long-term predictive models will have to be used to determine what, if anything, can be done to influence the effect. The scope and scale of these models will need to be far more extensive than those used for nuclear waste predictions, and the consequences of getting the predictions wrong would be far more serious. Thus, although being opposed to nuclear power, A. G. Milnes says in his preface:

It became clear to me that radioactive wastes constitute an environmental hazard which in no way overshadows the problems associated with many other waste products (CO_2, SO_2, heat, chemicals, heavy metals, etc.) whose management is haphazard to say the least, and whose 'disposal' has not yet led to widespread public concern. The positive side of the historical accident which has fixed such enormous attention on what may, in retrospect, seem to be an insignificant part of the whole process of environmental degradation is that it is laying a solid foundation for the better management of all other types of hazardous waste. (p. xiv)

However, nuclear power *is* unique in the way in which its risks are perceived, a point we touched on in Section 4.1. As you saw in Section 5.7.5, the political difficulties discussed in this section led the British government to abandon proposals for a high-level waste depository. So far, there is no organization responsible for high-level waste disposal in the UK. Furthermore, not one site in the world has yet been licensed for the large-scale disposal of nuclear waste, and it is unlikely that there will be a licensed site anywhere actually accepting high-level waste for some time to come: the present intention is that the vitrified waste already being produced will be stored near the surface for at least 50 years.

5.8.1 Ethics and the alternatives

At present, spent fuel in the UK is reprocessed and the waste then stored on or near the surface at Sellafield, eventually in a vitrified form. Some of those opposed to fuel reprocessing and high-level waste disposal suggest that storing the spent fuel at the reactor site where it is produced would reduce transport hazards, and could even represent a long-term solution. This procedure could have the advantage that it allows time for the wastes to decay, and for some acceptable disposal procedure to evolve. Blowers, Lowry and Solomon say of this idea:

The problem with this view is that it is not clear if the problems and uncertainties involving nuclear waste disposal, given the longevity of many of the radioisotopes, will ever be resolved; nor do they necessarily need to be since we do not live in a risk free world. All the major states that we have examined in fact favour disposal over long term storage. Existing disposal techniques can minimize safety risks, as is being tried in Sweden. Moreover, an ethical case can be made that the current nuclear waste problem should be 'solved' by those who created it (which is strongly argued in Sweden), without burdening future generations. (p. 318)

Here an ethical argument is injected into the debate about storage versus disposal. It revolves around the question of whether it is morally justifiable to subject future generations to risks and expenditure as a result of activities that do not benefit them. The implication is that disposal is the more responsible option because it does something to remove the burden of decision and action from those not yet born. Blowers, Lowry and Solomon also argue that if we plump for disposal, geological uncertainties impose on us an obligation to provide site monitoring and waste retrieval facilities for perhaps 100 years after the end of deposition at a site.

5.8.2 Should nuclear waste be reprocessed?

As Figure 5.12 shows, the handling of spent fuel need not involve reprocessing. Whether nuclear waste should be reprocessed or not is an important question, because reprocessing increases the volume of waste to be disposed of. It also results in the release of radioactivity into the environment (Figure 5.11) and exposes the workers to radiation doses.

One of the justifications for undertaking spent-fuel reprocessing is that the plutonium that is recovered can be used to fuel fast-breeder reactors. There appeared to be a need for these if nuclear power continued to expand at the rate suggested during the 1960s and early 1970s; indeed it was part of the case made for THORP (the thermal oxide reprocessing plant) at the 1977 Windscale inquiry. However, the time-scales for the introduction of fast-breeder reactors receded when the rate of growth of nuclear power slowed down, in the late 1970s and early 1980s. As a result, many fast-breeder reactor development programmes have now been cut back (as in the UK), postponed or simply cancelled. In addition, the reduction in growth rate of nuclear power internationally has meant that the anticipated shortages of uranium have not yet materialized.

If there is no current demand for the fissile material produced by fuel reprocessing, is there any advantage in undertaking the separation *now*? From the point of view of radiological safety, you have seen that there are advantages in delaying it. However, reprocessing does have some advantages; in the UK, for example, it concentrates the *high-level* wastes—which contain some 95% of the radioactivity—into a very small volume of solid glass (Figure 5.10), which is probably more leach-resistant than untreated spent fuel. Moreover, the investment has already been made in THORP, much of it having been paid for in advance orders for reprocessing by foreign countries such as Japan. To renege on these commitments would raise obvious political difficulties.

Let us, however, look at the costs from the standpoint of an electricity company with spent fuel on its hands. In a report prepared for Greenpeace (*THORP and the Economics of Reprocessing*, University of Sussex, 1990), Frans Berkhout and William Walker discuss the different options. They conclude that without reprocessing, there is a reduction in costs if long-term storage of spent fuel is adopted rather than disposal. Even if disposal is intended, the reprocessing does not change the economics significantly.

Berkhout and Walker argue further that economics cannot be the sole criterion in decisions on spent-fuel management. In particular, they are concerned about the accumulation of separated plutonium. In the case of overseas contracts, this will be returned to the country of origin if it is not turned into fresh reactor fuel. Berkhout and Walker maintain that the hazards posed by this stock of plutonium is a strong argument against reprocessing. We shall encounter this problem again in Chapter 8.

Summary of Section 5.8

1 The fact that there is now always strong local opposition to nuclear waste depositories of any kind has meant that the choice of sites has become primarily a political issue. As a consequence, sites cannot be chosen only on the basis of their geological suitability.

2   There are strong differences of opinion on whether high-level waste disposal can ever be undertaken with acceptable safety. The main factors giving rise to these doubts centre on whether it is possible to make predictions of radioisotope transport and geological stability over the very long time-scales involved.

3   The case for nuclear fuel reprocessing being part of waste management is weakened if there is not a serious commitment to a fast-breeder reactor programme.

4   Both economic and radiological arguments favour the delaying of reprocessing in countries where no facilities exist. However, in the UK, the large capital investments that have already been made in reprocessing and vitrification plant must be taken into account.

Activity 5.2

Read Extract 5.1.

(a)   In what way does the repository with which the extract is concerned differ from the type that we have concentrated on in the latter part of this chapter?

(b)   The hydraulic gradient for the Sellafield repository site was quoted in the extract to be 1 in 12 (0.083); this is well within the range appearing in Table 5.2. How does this value change when a possible repository at Dounreay is considered? If the Sellafield value replaces those cited in Table 5.2, how do the velocities of movement of water change in the four rock samples given in the table? Are any of them then consistent with the velocity of 'a few metres a year' cited in the extract?

divide gradient by ·083 then × answer by velocity in table.

Extract 5.1   From *The Independent on Sunday*, 19 April 1992.

Sellafield safety

Nuclear waste options narrow

Tom Wilkie examines the dilemma over disposal of radioactive material

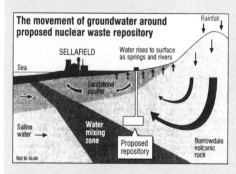

The nuclear industry decided in July last year to tip its waste into its own backyard. After abandoning potential sites in the south of England in 1987, it realised that only where nuclear power already had a presence could it hope to secure public acceptance to a waste repository.

This narrowed the choice to either Dounreay in the north of Scotland or Sellafield in Cumbria. Nearly two thirds of the waste that will go into

the repository is generated at Sellafield, and the industry's waste disposal company would have had to build new stretches of railway line if it had decided to transport the waste to Caithness. The monetary case was clear: it would cost about £2.5 bn to build and operate the repository at Sellafield, but nearly double that at Dounreay.

Only intermediate-level and low-level radioactive waste will be consigned to Nirex's repository. Low-level waste includes slightly contaminated protective clothing; intermediate-level waste includes water filters and discarded fuel-cladding hulls.

Highly radioactive waste from the reprocessing of spent nuclear fuel will remain stored on the surface at Sellafield for at least 90 years. The nuclear industry does not intend to start disposing of high-level waste until at least 2081.

At first, the geological case for Sellafield looked favourable. Because the rocks under Sellafield dipped down from the Lake District and out under the Irish sea, any underground water would move slowly—over a period of tens of thousands of years—from the repository straight on out under the sea, where if it eventually rose to the surface it would be diluted. The water in these basement rocks—known as the Borrowdale Volcanic Group—would be completely separate, the company thought, from that in the overlying sandstone cover.

Nirex has already spent tens of millions of pounds surveying the Sellafield site. It expects to submit a planning application later this year, to start excavating in 1995, and to put the first waste packages into the repository by 2005.

The results of its drilling programme, however, have revealed that there is a band of dense highly-saline water blocking the flow out to sea of the fresh water from the hills. Propelled by pressure from the head of water in the hills, this fresher water will "float" upwards on top of the saline into the rocks above, which have been tapped as a source of drinking water.

"Any sort of upward migration is what they do not want in that area and they have got it", according to John Mather, professor of geology at London University, and adviser to Cumbria County Council. The pathway for return to the environment is thus much shorter, and the speed with which the water is travelling is much faster than Nirex originally thought.

Professor John Knill, who is a specialist on deep underground water flows associated with reservoirs and tunnels, said: "The hydraulic gradient is about one in 12 upwards—that is not a low value. If you apply the formula that Nirex itself gives in its report, you get flow times of a few metres a year—these are not slow rates of water movement."

Harold Beale, the technical director for Nirex, said: "The water will come up under the sedimentary rocks. We have not yet deter-mined where the water goes. We believed it would come out further out to sea. But we do believe that the water system in the sandstone is decoupled from that in the basement rocks. It would be amazing if there were a watertight seal all the way."

The two-volume analysis of *The Geology and Hydrogeology of Sellafield* will form the basis for Nirex's safety case for presentation at the public inquiry. Copies of the report have already been sent to local authorities in the area and to the Department of the Environment and HM Inspectorate of Pollution.

But Mr Beale said that to get a full understanding of the site, "you have to go to depth and tunnel horizontally through the rock—yet you can't get down there and get the information without planning permission to dig the hole". That is the reason, he said, that Nirex intends to go to a planning inquiry without having completed its long-term safety case. If, when the case was complete by about 2001, there was doubt about the long-term safety of the repository, the Secretary of State for the Environment could then hold another public inquiry.

Professor Mather said Dounreay might be a better bet geologically and hydrogeologically. Because the terrain is largely flat, the ground-water flow there is expected to be sluggish—it will not be driven by a head of water from an elevated outcrop as is the case for the Sellafield Borrowdale Volcanics.

If there is any flow, he said, it is expected to be in the direction of the Pentland Firth "where there is a hell of a current". Any radioactivity would be so heavily diluted as to be virtually undetectable.

6 Reactor accidents and risks

Accidents always excite a lot of attention. Plane crashes, motorway pile-ups, pit disasters, explosions on North Sea oil platforms—all have a greater fascination than many people are willing to admit. In this chapter we are going to discuss accidents to nuclear reactors. We begin with a very brief general discussion of the dangers from nuclear reactors, and then we discuss three of the most important reactor accidents that have occurred.

Luckily, however, reactor accidents involving the release of radioactivity are very rare. Consequently, the prediction of risks from nuclear power based on *experience* is difficult. The last part of the chapter is therefore concerned with how the risks from accidents can be *estimated*, and with the limitations of those estimates.

From *The Guardian*, 27 February 1992.

6.1 Radioisotopes and reactor accidents

In Chapters 4 and 5 we established that the principal danger from nuclear power arises from the radioactive wastes in the nuclear fuel cycle. The risk to humans arises both from radioactivity *outside* the body (giving an *external* radiation dose) and from radioactivity *entering* the body, either through eating, drinking or breathing (giving an *internal* radiation dose).

But exactly how serious is the danger from operating nuclear reactors? One way to obtain a feeling for the magnitude of the potential problem is to compare the amounts of radioactive fission products in a reactor core with the maximum concentrations in air and water recommended for the general public. The most dangerous isotopes in the core are the isotopes of plutonium, and the fission products ^{137}Cs, ^{131}I and ^{90}Sr. We shall make the point by using one simple example. The isotope ^{90}Sr, if ingested, can become incorporated into bone as a consequence of its chemical resemblance to calcium: they are both in Group II of the Periodic Table. The maximum permitted activity of ^{90}Sr in one cubic metre of drinking water is 1.1×10^4 Bq, or, expressed as a concentration, 1.1×10^4 Bq m^{-3}. Now the activity of the ^{90}Sr in the core of a single PWR whose fuel is about to be replaced is about 2.9×10^{17} Bq. Suppose the PWR core were to be suddenly immersed in a very large quantity of water, and all the ^{90}Sr were dissolved.

▷ What volume of water would have to be present if the final concentration of dissolved ^{90}Sr is to be equal to the maximum permitted value for drinking water?

▶ If the volume is V, then the maximum permitted activity of $1.1 \times 10^4\,\text{Bq m}^{-3}$ is given by

$$\frac{2.9 \times 10^{17}\,\text{Bq}}{V}$$

Hence

$$V = \frac{2.9 \times 10^{17}\,\text{Bq}}{1.1 \times 10^4\,\text{Bq m}^{-3}} = 2.6 \times 10^{13}\,\text{m}^3$$

This volume of water is very large. It would occupy a lake about 2 m deep and the area of Australia! What this shows is that the quantities of the radioisotopes present in the core of a reactor are *enormous* compared to the amounts permitted in a cubic metre of water. (The same observation would apply to the possible concentration of reactor radioisotopes in air compared to the permitted concentrations.) Hence, if the radioisotopes were to escape, they would have to be very widely dispersed or diluted before their concentrations in air or water were at what is considered to be a safe level. Thus, when there is an accident in a nuclear reactor, it is the release into the environment of the radioactive fission products contained in the reactor core which represents the danger to the public. There have been three major accidents in which radioactivity in a reactor core has been released, namely at Windscale (UK), Three Mile Island (USA) and Chernobyl (former USSR). We shall discuss each in turn to see what lessons can be learnt from them.

6.2 The Windscale reactor accident

The Windscale accident, in 1957, involved a fire in a relatively low power (180 MW) reactor used for producing plutonium for nuclear weapons. The design of the Windscale reactor was totally different from that of any reactor used for electricity production, and the lessons that can be learnt from it are very limited. Nevertheless, the accident is the most serious to have occurred to any reactor in the UK, and, for this reason, it is important that you know about it.

The fundamental reasons for the accident are to be found in the graphite moderator used in the Windscale reactor. The normal crystal structure of graphite is shown in Figure 6.1. When the fast neutrons produced during fission pass on their energy to the graphite, they collide with and displace carbon atoms from their usual positions. At high temperatures the transferred energy is soon dispersed through the graphite, and the displaced atoms settle down into a structure just like that in Figure 6.1, but with slightly different interatomic distances. However, if the graphite temperature is low (less than 300 °C), some of the absorbed energy results in a *distortion* of the structure: following collisions with fast neutrons, some carbon atoms are displaced to, and remain in, positions quite different from those shown in Figure 6.1, and the energy that brings this about is then stored in the graphite structure as strain energy. This is analogous to the energy stored in a stretched piece of elastic, where atoms are also displaced from their equilibrium positions. In the case of a piece of elastic, the stored energy can be released by letting go of it. In the case of graphite, increasing its temperature allows the atoms to return to their original, equilibrium positions, and the stored strain energy is then released as heat energy. The energy stored and released in this way is often called 'Wigner energy', after Eugene Wigner, the Hungarian-born physicist who first explained it.

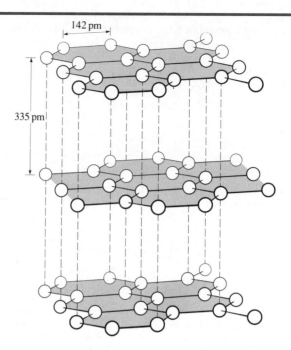

Figure 6.1 The normal structure of graphite with the carbon atoms in their mean equilibrium positions: hexagons of carbon atoms share edges to form flat sheets, which lie parallel to one another. 1 pm = 10^{-12} m.

6.2.1 The sequence of events

There were two identical reactors at the Windscale site, which have long since been shut down. Each had a graphite moderator and natural uranium metal fuel, clad with aluminium. The coolant was air, and this was simply drawn into the reactor by large fans, and then discharged through 125 m-high chimneys into the atmosphere (Figure 6.2); no electricity was generated. At the insistence of Sir John Cockcroft (a prominent nuclear physicist who became Chairman of the United Kingdom Atomic Energy Authority), each chimney had a filter at the top; these filters—which were very prominent (Figure 6.3)—were known somewhat irreverently by the workers as 'Cockcroft's Folly'. His wisdom was only appreciated later.

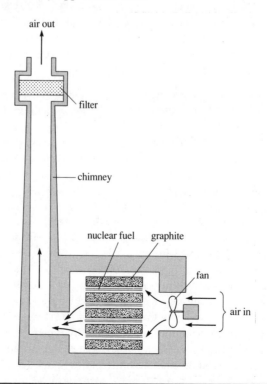

Figure 6.2 A cross-section through the Windscale reactor.

▷ How is this reactor different from gas-cooled power reactors like the AGR?

▶ In gas-cooled power reactors the coolant is either carbon dioxide or helium, and it circulates in a closed loop (Figure 3.9).

During normal operation, the graphite temperature was comparatively low (150 °C), and the Wigner energy stored in the graphite built up. However, it was known that if the energy stored in this way built up too much, it could be released spontaneously, heating up the graphite in an uncontrolled way. In fact, this had happened in 1952 without any serious consequences. To prevent spontaneous release occurring, the strain energy was therefore released from time to time in a controlled way, before it had built up to a dangerous level.

To have a controlled release of the stored energy, the reactor was operated without any air flow; that is, the cooling was turned off. This allowed the energy produced in the fuel to heat up the surrounding graphite moderator, thereby releasing the stored Wigner energy. Once this energy release was over, the fans for circulating the cooling air were then turned on again.

Although the air-circulating system was turned off during the release of Wigner energy, air (and hence oxygen) remained in the reactor. Graphite does not burn as readily as coal, but it does spontaneously ignite in air if the temperature is greater than about 700 °C. Uranium metal also ignites spontaneously in air at around this temperature.

Normally, when the air supply was turned on again after a controlled Wigner energy release, the temperature inside the reactor dropped. On 1 October 1957, however, although the temperature of most of the core dropped, in one part it *rose*. After several hours, a release of radioactivity from the chimneys alerted people to the fact that something was seriously wrong. By that time, parts of the core were red hot, reaching a temperature of 1 300 °C. Actual flames were seen by workers who peered into the core through ducts used for loading and unloading fuel.

Figure 6.3 Photograph of the chimney of the Windscale reactor.

What had happened? The Wigner energy release had not taken place properly over the whole reactor core, and the fuel and cladding had overheated in one region. The cladding on some of the fuel elements had then melted, and both the uranium and the graphite had caught fire. Hence, putting on the air cooling fans at what was thought to be the end of the Wigner energy release simply made things worse, by fanning the fire! The fire was eventually put out by pumping water into the reactor.

Although the filters on the chimneys very significantly reduced the quantities of radioactive fission products released, they were not effective in preventing the release of volatile isotopes like ^{131}I, or the noble gases, like ^{85}Kr. In addition, the volatile isotope polonium-210, an α-particle emitter (half-life 138 days), was also released. This is not a fission product, but was being produced in the reactor for weapons use (we shall discuss this application in Chapter 8).

▷ Why are α-particle emitters especially dangerous?

▶ α-particle emitters are more dangerous than other radioisotopes because the α-particles are densely ionizing and cause intense radiation damage to living cells (Section 4.2).

6.2.2 The implications of the Windscale accident

Following the accident, milk supplies from an area of 500 km^2 around the reactor were destroyed, since this was the most direct route by which ^{131}I could enter the human food chain. You saw earlier (Section 4.3) that iodine accumulates in the thyroid gland, and radioactive iodine can cause thyroid tumours. None of the workers were exposed to the levels of radiation which produce radiation sickness (greater than about 2 Sv; see Table 4.2), although many received radiation doses far in excess of permitted levels, as did some local people.

A full report of the accident was not made public until 1982, even though both the Ministry of Defence and the UK Atomic Energy Authority were in favour of releasing the report prepared at the time of the accident. In the 1982 report, prepared by the National Radiological Protection Board, it was estimated that the release would have produced 13 additional cancer deaths. Following criticisms of the report, this estimate of the number of cancer deaths was later increased to 32, most of the increase being due to the effects of the polonium-210.

The accident showed the importance of not running reactors with a graphite moderator at low temperatures. Graphite-moderated reactors for producing nuclear power run at temperatures higher than that at which Wigner energy is stored by the graphite.

Summary of Sections 6.1 and 6.2

1   The principal threat to the public from an accident in a nuclear reactor is the threat of the release of the fission product radioactivity in the core into the environment.

2   The Windscale accident of 1957 occurred because the reactor normally operated at only 150 °C, a temperature at which strain energy built up in the graphite moderator.

3   Routine release of strain energy was achieved by the increase in temperature which occurred when the fan-driven air-cooling was stopped. On one occasion this was not fully effective. The energy released caused an excessive rise in temperature, and a subsequent fire in the fuel and graphite; the fire was eventually put out with water.

4   Substantial amounts of volatile radioisotopes were released, which were estimated to have been responsible for 32 additional cancer deaths.

6.3 The accident at Three Mile Island

The accident at Three Mile Island (Harrisburg, Pennsylvania, USA) on 28 March 1979 occurred at one of two PWRs on the site, which had been built by Babcock and Wilcox (USA). The reactor concerned, No.2, first started operation in December 1978, only three months before the accident occurred. As a result, the amount of fission products in the core of the reactor was not very large.

To understand the accident, you should remember the following features of a PWR: the uranium dioxide fuel is clad in a zirconium alloy; light water acts as both coolant and moderator; the water is prevented from boiling by pressurizing it; the steam cycle is indirect; the reactor is in a special containment building. You can remind yourself of these details by returning to Figure 3.10 and Section 3.2.2.

6.3.1 The accident

The following description of the accident relies on many references to the lettered items in Figure 6.4 (a photograph of the reactor is shown as Figure 6.5). The key points that started the accident sequence, and the time-scales involved, were as follows:

1   The reactor was running at nearly full power when a pump (A) circulating cooling water from the condenser (B) to a steam generator (C) stopped operating.

▷   What happens to the temperature of the primary coolant if the steam generator becomes less good at removing heat from it?

▶   The primary coolant will heat up.

2   Some 3–6 seconds later, the pressure relief valve (D) in the pressurizer (E) opened, because the increase in temperature of the primary coolant was accompanied by an increase in pressure in the cooling circuit.

3   After 9–12 seconds the reactor shutdown rods went in automatically, because the coolant pressure had exceeded its preset limit. This stopped the chain reaction.

▷   Does stopping the chain reaction stop heat production in the core?

▶   No; heat is still produced by the radioactive decay of the fission products in the fuel (Section 3.1.2).

4   After 13 seconds the coolant pressure and temperature dropped, but *the pressure relief valve (D) failed to close.*

This failure was the key to everything that followed. Its importance was that the open valve allowed water to escape from the reactor cooling circuit, initially into a special overflow tank (F).

After this, water continued to flow out of the primary cooling circuit. When the special overflow tank was full, it overflowed into the sump (G), from where it was pumped into a water storage tank (H) in an adjacent auxiliary building (I). From this building—which is separate from the containment building (P)—gaseous fission products (which were not removed by the filters (J) in the chimney) were discharged into the environment; the largest release was of ^{85}Kr.

5   After 2 minutes, the falling coolant pressure automatically triggered the emergency core cooling system (K). However, the operators believed wrongly that there was too *much* water in the core, and after 3 minutes of operation they turned off the emergency cooling system. An auxiliary pump (L) in the steam generator circuit was

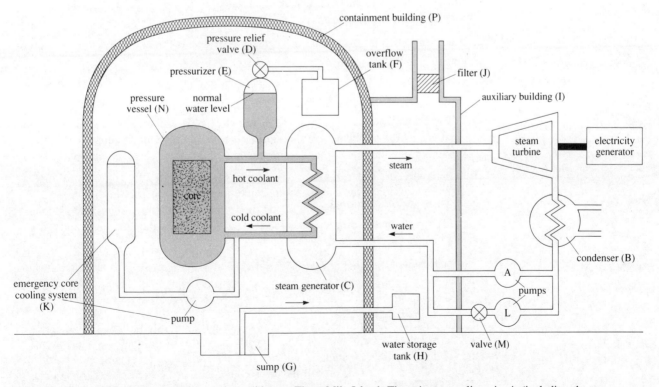

Figure 6.4   Key PWR features relating to the accident at Three Mile Island. The primary cooling circuit (including the pressurizer) is shown in colour.

Figure 6.5 The Three Mile Island reactors. The accident occurred in reactor No. 2, which is indicated by the arrow.

switched on, but was ineffective because a valve in the line (M) had been left closed by mistake after maintenance.

All this time the operators, misled by faulty instrumentation, struggled to understand what was happening.

▷ What happens to the primary coolant if the pressure drops?

▶ The primary coolant will boil if the pressure drops to the point at which the temperature of the cooling water is higher than its boiling temperature.

6 Eventually, the water in the primary cooling circuit started to boil, because the pressure had dropped too much. At the same time, the water level in the pressure vessel (N) dropped below the top of the fuel rods; without any cooling, the cladding started to melt, and a chemical reaction took place between the zirconium and the steam, producing hydrogen:

$$Zr(l) + 2H_2O(g) \longrightarrow ZrO_2(s) + 2H_2(g) \tag{6.1}$$

The hydrogen collected in the top of the pressure vessel, making it difficult to refill with coolant.

7 Adequate cooling of the whole core was eventually re-established 16 hours after the start of the accident, but not before nearly one-third of the fuel had melted, and the inside of the reactor containment building (P) had become contaminated with fission products. The reactor will never operate again, and the cost of decontaminating the buildings and dismantling the reactor has been estimated at US $1 billion (1988 prices).

Activity 6.1

Now go back over the sequence of events which contributed to the accident, and identify examples of (a) equipment failure, (b) inadequate maintenance checks and (c) operator error.

Two quotations from the Report of the President's Commission on the accident, published in 1979, will illustrate its tone.

During the first few minutes of the accident, more than 100 alarms went off, and there was no system for suppressing the unimportant signals so that the operators could concentrate on the important alarms. (p. 11)

... it is our conclusion that the training of the Three Mile Island operators was greatly deficient. While training may have been adequate for the operation of the plant under normal circumstances, insufficient attention was paid to possible serious accidents. And the depth of understanding, even of the senior reactor operators, left them unprepared to deal with something as confusing as the circumstances in which they found themselves. (p. 10)

Although there is no doubt whatsoever that this was a very serious, and costly accident, it was *not* serious from the point of view of environmental releases and consequent health effects. Thus, although about one-third of the fuel melted, you can see from Table 6.1 that noble gas isotopes (mainly ^{85}Kr) were the only radioisotopes to escape into the environment in significant quantities. As a consequence of this small release, the radiation dose to the public was equally small. The average radiation dose to anyone living within 15 km of the site was estimated to be 0.08 mSv, although three of the reactor operating staff received doses of between 30 and 40 mSv. The Report of the President's Commission on the accident concluded that:

It is entirely possible that not a single extra cancer death will result. And for all our estimates, it is practically certain that the additional number of cancer deaths will be less than 10. (p. 12)

To put these figures into perspective they observed that there would be about 325 000 cancer deaths from other causes among the population within 80 km of Harrisburg.

Table 6.1 Percentages of some of the biologically important fission products in the core which entered the environment from the Three Mile Island and Chernobyl accidents. Notice that the two columns relevant to Three Mile Island show that most of the fission products leaving the core stayed within the containment building.

Element	Three Mile Island		Chernobyl
	Percentage of total core content for the element concerned		
	leaving the core	entering the environment	entering the environment
noble gases (Kr and Xe)	48	1	100
iodine	25	3×10^{-5}	20
caesium	53	not detected	10–13
ruthenium	0.5	not detected	3
lanthanides	nil	nil	2.5

The other health effect that the President's Commission noted was mental stress, which they said was short lived, and most prevalent in families with young children living within 8 km of Three Mile Island. This is obviously an important consequence, but one that is difficult to assess in terms of long-term health effects.

6.3.2 The implications of the Three Mile Island accident

The importance of the 'human factor' in determining the behaviour of engineering equipment was highlighted by the President's Commission. One of the many problems in coping with *rare* accidents is that of *operator inexperience*; if something only happens once every few years, those concerned may forget how to respond. One way round this problem is to use reactor simulators. These take the form of a control console similar to, or identical with, that of the power station concerned, but with the controls and gauges connected to a computer, which is programmed to *simulate* the behaviour of the reactor and associated power plant. The reactor operators can then be trained to respond to a whole range of different situations, and this training can be repeated at regular intervals. This was one of the recommendations of the President's Commission, and is standard practice in the UK.

A second factor in reducing the scope for human error is to make it easier for the operator to respond by careful choice of the instrumentation and of the control room layout. A third way is simply to have as much of the response as possible determined by computers—that is, to dispense with humans! Indeed, in the UK the current philosophy for the most recent reactors is that routine reactor control is by computer, and that for the first half-hour after an incident leading to a shutdown there should be no operator intervention, control being by computer only (Box 6.1). The reasoning behind this is:

(i) the computer programs are written and tested by a team of people in an atmosphere well removed from that of an emergency;

(ii) the programs can incorporate factors relevant to the whole plant which it would not be reasonable to expect an operator to remember, or to be familiar with;

(iii) the problem of computer failure can be overcome by using several computers, and by cross-checking the results from two or more computers.

6.1

Box 6.1 An aside on the use of computers

The idea that using computers should make things safer is not always accepted. Some people still mistrust them, and can support their prejudices with stories of the mistakes that 'computers' have made—usually involving bills or banks. However, it is usually not the computers that have made the mistake, but their *operators*. The phrase 'junk in, junk out' is often used for these mistakes; that is, if you feed the wrong information into a computer program, it will give you the wrong answer!

In fact, computers are already so widely used in situations where their reliability is essential for safety, such as in space vehicles, that it does not make sense to question their use *in principle*. Nevertheless, one clearly has to look at each different application in detail, to ensure that the limitations of their use are appreciated.

Of course, mistakes do occur as a result of errors in the computer programs, and very considerable effort is being devoted to try to develop 'error free' programming systems. In addition, computer programs may become so complex that they are difficult to check properly. But this does not invalidate their use; in the majority of applications, for example flying aircraft, controlling CD players or running supermarket checkouts, they work extraordinarily quickly, reliably and accurately day in and day out. They have the overriding advantage that they do not suffer from fatigue, boredom, or the effects of a good night out or, indeed, from any of the factors that can affect human judgement. And their memory recall is superb! ∎

The Three Mile Island accident served to focus attention on a range of unsatisfactory aspects of the US reactor scene, and to cause greater scrutiny to be given to practices world wide. For some people it also served to demonstrate that nuclear power was simply unacceptable. For example, Peter Bunyard, writing in *The Ecologist* (3 June 1979), concluded his article on the accident by saying:

> *Surely the lesson of Harrisburg is obvious. We cannot afford nuclear power: it costs too much, its wastes cannot be contained, its safety defies prediction and it undermines society.*

An alternative view is set out by Professor S. E. Hunt, in his book *Nuclear Physics for Engineers and Scientists* (Ellis Horwood, 1987, p. 361):

> *On the positive side, despite one defect in the apparatus and three major operator errors, the defence in depth safety measures worked to the extent of limiting the health hazard to the general public to a negligible level.*

Summary of Section 6.3

1   The 1979 accident at Three Mile Island began with rising coolant temperature and pressure caused by the failure of a pump for feeding water to the steam generator.

2   The resultant rise in temperature and pressure automatically opened a pressure relief valve in the primary coolant circuit and shut down the reactor.

3   The pressure relief valve failed to close when the coolant pressure decreased, and water was continuously lost from the primary cooling circuit. This loss of water became more damaging when the operators misread the situation and turned off the automatically triggered emergency core cooling system. The level of the coolant eventually fell below the top of the fuel rods.

4   Heat from fission product decay caused the dwindling coolant to boil and the fuel to partially melt. The zirconium alloy cladding reacted with steam, producing hydrogen gas.

5   Coolant containing fission products flooded the basement of the containment building, but very little radioactivity was released to the environment.

6.4 The Chernobyl accident

The accident to the Chernobyl reactor on 26 April 1986 is the most serious, and the most dramatic, which has ever occurred to a nuclear reactor. At the time, the affairs of the USSR were still cloaked in secrecy. The fact that the accident had occurred at all only came to light when raised radiation levels were detected at a Swedish nuclear power station (Forsmark), which could not be explained by any local release of radioactivity.

Subsequently, pictures of the shattered reactor building and of the aftermath of the accident have been shown world wide. The heroic attempts of the workers, the firefighters and the armed forces to control the fires and to ensure that the reactor was made safe have featured in numerous television documentaries, as have analyses of the effects of the accident. Indeed, the breadth and often emotive nature of the coverage has been such that it is difficult to approach consideration of the accident with an open mind.

6.4.1 Some background to the accident

The feature of the RBMK reactor which is most relevant to the Chernobyl accident is the *combination* of a light-water coolant with a graphite moderator. You can remind yourself of this and other features of the reactor by returning to Figure 3.11 and Section 3.2.3. Notice that the reactor uses a direct steam cycle. The pressurized water coolant normally enters the bottom of the core at 270 °C, and leaves the top at 285 °C, when it consists of 85% water and 15% steam. The steam is then passed to a turbine, and the water is recycled to the core.

When discussing the components of nuclear reactors in Section 3.1.1, we implied that in light-water-cooled systems, the ^{1}H atoms in H_2O absorb significant numbers of neutrons. This absorption by the coolant therefore supplements the neutron absorption of the control rods, and the combined absorption holds the neutron population and the power level constant when the reactor is critical.

If there is a reduction in the amount of light water near a fuel element—for example, if a steam bubble (a void), in which the density of H_2O molecules is necessarily low, is formed—then neutron absorption in the water will be reduced. If the reactor were critical before the steam bubble formed, the presence of the bubble would make it supercritical if this decreased neutron absorption were the only effect. There are, however, other factors involved. For example, if the amount of moderator is reduced, more neutrons would be absorbed by the ^{238}U in the fuel; this would counteract the effect of decreased neutron absorption by the light-water coolant.

In practice, whether the presence of a void makes the reactor power increase or decrease depends on the amount of fuel, its degree of enrichment and its arrangement in the moderator. In the case of the Chernobyl reactor, however, the effect of a void in the coolant was to increase the rate at which fission occurred: the reactor was said to have a **positive void coefficient**.

Let us consider the implication of this. If boiling occurs in the coolant of a reactor with a positive void coefficient, the power increases. The increase in power then results in more steam (hence more voids) being produced, which, in turn, produces even more power; there is thus a potentially very dangerous and unstable situation.

We say *potentially* very unstable, because the control rods can be used to counteract this effect. Indeed, one of the most successful reactor types in the world, the Canadian CANDU reactor discussed in Activity 3.1, has a positive void coefficient. However, in addition to their normal control systems, CANDU reactors have two completely independent shutdown systems; in RBMK reactors the control and shutdown systems are combined, and there are no alternative systems.

6.4.2 The effect of fission products

So far, we have talked about fission products in terms of the potential hazard they represent to humans. However, some of them also represent a hazard to the reactor itself, in so far as they or their daughter products absorb neutrons. The most important neutron absorber produced in this way is xenon-135, which is produced from a fission product, tellurium-135 (half-life less than 1 minute), via iodine-135 (half-life 6.7 hours), following two β-decays:

$$^{135}_{52}\text{Te} \xrightarrow{\beta\text{-decay}} {}^{135}_{53}\text{I} \xrightarrow{\beta\text{-decay}} {}^{135}_{54}\text{Xe} \tag{6.2}$$

When ^{135}Xe acts as a neutron absorber, it simply forms the stable isotope, ^{136}Xe.

The amount of ^{135}Xe produced when a reactor is running steadily depends on the power level at which the reactor is operating—the higher the power, the more ^{135}Xe there is.

▷   If the reactor is shut down, will ^{135}Xe production stop immediately?

▸   No; ^{135}Xe will be produced until all the ^{135}Te and ^{135}I that has already been formed has decayed.

Indeed, following any sudden drop in power, the amount of ^{135}Xe present actually *rises* initially. This may seem surprising, but it happens because there are fewer neutrons in the core when the power drops, and there is therefore less chance of ^{135}Xe being transformed into ^{136}Xe by absorbing a neutron. Following a drop in power, the amount of ^{135}Xe in the core only starts to fall after about 12 hours (Figure 6.6).

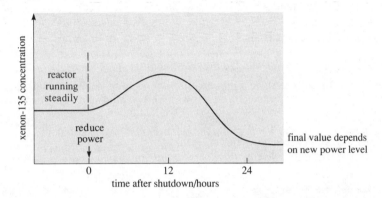

Figure 6.6   Change of ^{135}Xe concentration in a reactor following a reduction in power.

Since ^{135}Xe is such a strong neutron absorber, a rise in its concentration will reduce the neutron population, and make the reactor subcritical. This phenomenon is known as *xenon poisoning*.

▷   How can the reactor be made critical again if this happens?

▸   The reactor can be made critical again by raising the control rods.

With this additional background we can now consider the accident at Chernobyl.

6.4.3 The sequence of events

The Chernobyl accident occurred in one of four RBMK-1000 reactors on the site, which is 100 km north of Kiev; the nearest large town is Pripyat (population about 49 000), about 8 km away. The reactor, which produced 1 000 MW of electricity, had been operating since 1984, and those on the Chernobyl site were the most successful in the USSR, having averaged 83% of their maximum output during 1985. This performance ranked them among the top 25% of reactors world wide. As we discussed earlier, the RBMK reactor is graphite moderated and light-water cooled. Its fuel elements consist of enriched UO_2 clad in a zirconium alloy, and surrounded by the zirconium alloy pressure tubes.

In contrast to the Three Mile Island accident, the one at Chernobyl occurred during a test designed to try to find out how long the electricity generators would continue to generate electricity properly if the steam supply to the turbine was cut off: the turbine continues to rotate and generate electricity for a while as it runs down. This type of test formed part of a safety study, and had been conducted several times before.

An important characteristic of this reactor is that its void coefficient is large and positive if it is operated at less than 20% of full power. For this reason, it was planned to conduct the test at about 25% of full power.

When the operators started to reduce power in preparation for the test, this reduction was accompanied by an increase in the ^{135}Xe concentration (as we discussed earlier), requiring the control rods to be withdrawn in order to maintain criticality.

Before the test power was reached, an order came to delay the test, and to continue supplying electricity at about 50% of full output. During this time the ^{135}Xe concentration continued to rise, threatening a further reduction in neutron population. Thus, the control rods had to be withdrawn *even more* in order to keep the reactor critical. The result was that, when the time came to conduct the test, the control rods were so far out of the reactor that control was difficult. Indeed, when the operators eventually tried to lower the power for the test by slight re-insertion of the control rods, it fell uncontrollably to about 7%, a power level at which the reactor was known to have a positive void coefficient, and hence to be potentially unstable.

This drop in power also initiated a *further* rise in the ^{135}Xe concentration. Under these abnormal conditions the reactor should have shut itself down automatically, but the operators prevented this happening. This was one of several violations of their operating instructions. Despite this, and despite the fact that the reactor was operating well below the planned power level, it was decided that the test should go ahead—an extraordinary decision. The steam supply to the turbine was therefore shut off.

At this stage there were only about 7 control rods in the core, compared to the minimum of 30 specified in the operating instructions (the reactor has a total of 211 shutdown and control rods), and *all* of these 7 control rods were relatively far out of the core. This was another violation of the operating instructions. As a result, when there was an increase in the steam produced in the core, and the positive void coefficient then made the reactor supercritical, the control rods could not be moved into the core quickly enough to restore the reactor to a critical condition.

The order was then given to shut down the reactor, but the shutdown rods could not be moved quickly enough to be effective, requiring 18 seconds to go from the fully-out to the fully-in position. The net result was that the reactor became increasingly supercritical, and in 3–4 seconds the power rose to *100 times* its maximum design value. In a situation like this, 'reactor shut down' is hardly the right expression, since this implies something very orderly; in fact much of the core melted and dispersed.

Although large amounts of nuclear energy were released, the explosion that ruptured the pressure tubes and lifted the 2 000 t reactor cap off arose from *chemical interactions* between the molten fuel and the cooling water; the explosive energy from this interaction was estimated to be 0.5–1.0 GJ, equivalent to 100–200 kg of TNT. (About 1 kg is all that is needed to disable a jumbo jet.) In addition, there were also chemical interactions between:

(i)   the zirconium alloy cladding and the coolant (generating hydrogen)—see Equation 6.1;

(ii) the graphite moderator and steam (generating hydrogen and carbon monoxide);

(iii) hydrogen, air, and carbon monoxide (generating carbon dioxide and steam);

(iv) the graphite moderator and air (generating oxides of carbon).

The graphite caught fire, and 10% of it was burnt. The whole situation was so complex that a complete picture of the full accident sequence may never be obtained.

Since the building containing the reactor was only a light structure, it was simply blown apart by the explosion (Figure 6.7). Table 6.1 (p. 127) shows the resulting releases of radioisotopes, and compares them with those from Three Mile Island. From these data it is clear that the Chernobyl accident was very severe, both in terms of damage to the reactor *and* the radioactive releases.

Figure 6.7 The Chernobyl reactor buildings after the accident.

6.4.4 The distribution of radioactivity

Because of the explosion and the very large amount of energy released, the fission products were carried high up into the atmosphere on a plume of hot gas, and this continued for the ten days that it took to bring the fires and the releases of radio-activity under control. The pattern of dispersal of the radioisotopes during this time depended on the weather—notably on the wind direction and the rainfall (since the fission products tend to be carried down from the clouds with rain). The dispersal pattern is shown in Figure 6.8. You can see from this figure that the wind was blowing towards the north for the first few days after the explosion, which is why the radioactive cloud was first detected in Sweden. After that, the winds changed, and in the six-day period after the accident, the radioactive cloud moved in nearly every direction.

The way in which the radiation dose had built up across Europe after seven days is shown in Figure 6.9.

▷   Why, in Figure 6.9, do you think there is a region in the UK which received a higher level of radiation dose than the rest of the country?

▶   Although the radioactive cloud moved all over Europe, the main deposition onto the ground below occurred with rainfall. The region in the UK with the higher doses (north Wales, the Lake District, south-west Scotland) experienced a high rainfall when the radioactive cloud was overhead.

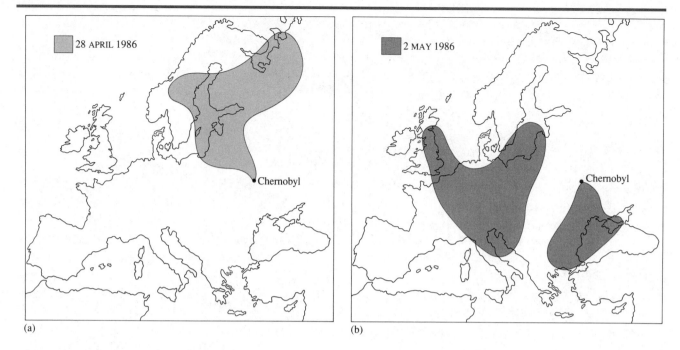

Figure 6.8 The position of the radioactive cloud from the Chernobyl accident, (a) 2 days, and (b) 6 days after the accident on 26 April 1986.

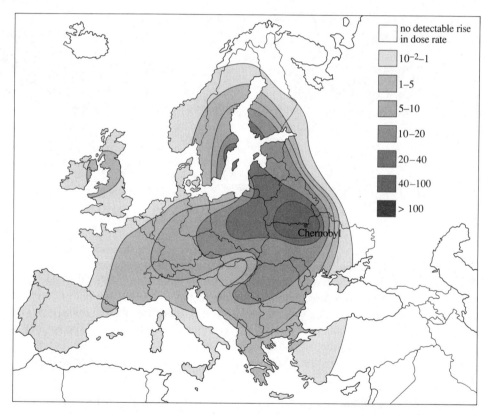

Figure 6.9 The pattern of the increased radiation dose across Europe on 3 May 1986. Notice that the increase is measured in multiples of the normal background dose rate.

As you saw in Section 6.1, the greatest radiological hazards are from ^{137}Cs, ^{131}I, ^{90}Sr and the isotopes of plutonium. ^{131}I is concentrated in milk, but, because its half-life is only 8 days, it does not represent a long-term contamination hazard. Control of exposure to ^{131}I therefore involves avoiding the use of contaminated food products; such control is particularly important for children, who are most at risk.

The long-term hazards arise from plutonium and the radioactive isotopes of caesium and strontium. The dispersal pattern will not be the same for all radioisotopes, since it depends on the *form* of the isotope (for example its particle size or solubility in rain-water) as well as on the weather pattern. The dispersal of the plutonium isotopes was restricted to an area within a 30 km radius round Pripyat, but the ^{137}Cs was dispersed much more widely, significant contamination occurring up to 300 km from Chernobyl. (It was ^{137}Cs which accounted for the main radiation doses in the UK.) The distribution of ^{90}Sr was intermediate between that of plutonium and caesium.

6.4.5 The radiation doses and their implications

The radiation doses to workers attempting to control the fires and the reactor were extremely high. Over 200 suffered from radiation sickness, and 31 died, mostly from radiation exposure. In the days following the accident, 46 000 people were evacuated from the region within a radius of 10 km around Pripyat. Subsequently, a further 90 000 people were evacuated from a 30 km radius exclusion zone around Pripyat. In addition, there were widespread restrictions on the use of fruit, vegetables and animal products (notably milk) from the contaminated regions.

Evacuation of the population in the outer, 30 km, zone was said to have limited the dose to the majority of the population in this region to less than 250 mSv in the first few months following the explosion, although some people were estimated to have received doses of between 300 and 400 mSv.

In Section 4.6, we said that high doses were represented by *instantaneous* doses in excess of 100 mSv. Here the dose was about 250 mSv, but it was received over several days. This is therefore a borderline case, but the period over which the dose was absorbed makes it reasonable to assign it to the 'low dose' category.

▷ Assuming that this is the case, use the data in Section 4.6 to estimate the probability that the people exposed to 250 mSv (0.25 Sv) will die of cancer as a result of their exposure.

▸ As the population is a mixed one (including children), the low-dose risk factor to use is a 5% chance of developing a fatal cancer per sievert. The risk of death for those exposed is thus 0.25 Sv × 5% chance of death—that is, a 1.25% risk of death. This is 1.25 in 100, or 1 in 80.

This sounds quite high, so let us try to put this risk in perspective. In the UK, we have already said (Section 4.7) that the risk of dying from cancer is 20% (1 in 5). Thus, if this UK figure for cancer incidence applies to the former USSR, then the accident at Chernobyl increased the risk of cancer in those receiving a dose of 250 mSv by (1.25%/20%)—that is, a factor of 1/16th, or 6.25%. Thus, for this group of people (that is, the majority of the population in the outer zone) the radiation dose from Chernobyl increased the risk of dying from cancer by 'only' 6.25%; in fact from 1 in 5 to about 1 in 4.7. Most of this group will die from other causes unrelated to the Chernobyl episode.

Evacuating the population from the 30 km zone around Pripyat also reduced the risks arising from plutonium-derived activity. However, the ^{137}Cs was so widely distributed that the population at risk from it is large. Quite *how* large depends on what level of risk, and hence of exposure, is assumed. An easily determined index of exposure is the activity per square metre at the Earth's surface. Using ground contamination values for ^{137}Cs of >185 kBq m^{-2}, the International Atomic Energy Agency (IAEA) quote an exposed population in the three republics concerned (Byelorussia, Russia and the Ukraine) as 825 000 (*The Chernobyl Project: an Overview*, IAEA, 1991). On the other hand, using a lower value of 37 kBq m^{-2} for ^{137}Cs, the figure for the exposed population rises as high as 4 million (*The Chernobyl Legacy*, Friends of the Earth briefing sheet, 1991). (For comparison, the ground levels of ^{137}Cs in Cumbria and North Wales from the Chernobyl accident were 20–25 kBq m^{-2}.)

The populations living in the contaminated areas are naturally very concerned about the long-term health consequences of their exposure. As a result, in 1990, the Soviet authorities decided that anyone expected to receive a lifetime dose of greater than 350 mSv should be considered for permanent relocation.

The team organized by the IAEA looked at many aspects of the Soviet investigations, both checking their methodology and undertaking independent investigations. Their results suggest that the Soviet authorities have, if anything, *overestimated* the risks. An example of this is shown in the comparisons for the estimated dose *over 70 years* to a group of settlements in the contaminated region (Table 6.2); these were outside the 30 km exclusion zone, but in regions where the ^{137}Cs ground levels exceeded 555 kBq m^{-2}. In Table 6.2, knowledge of the ground levels of the different radio-isotopes has been used to calculate the average radiation doses to individuals in the contaminated regions.

Table 6.2   A comparison of two estimates of the average radiation dose over 70 years to individuals from the same exposed populations in the former USSR.

	Total 70-year dose	
	Soviet estimate/mSv	IAEA estimate/mSv
external dose	80–160	60–130
internal dose	60–230	20–30
total dose	140–390	80–160

Indeed, the IAEA report is critical of the basis used by the Soviet authorities for relocation. It argues that the resources diverted to relocation were larger than were justified by the radiological dangers. The anxiety and stress occasioned by the accident extended far beyond the contaminated areas, and the report suggested that much of the expenditure was a political attempt to alleviate this, rather than a considered effort to cut radiation exposure.

One of the general conclusions on the health impact of the accident in the IAEA study was:

> *There were significant non-radiation-related health disorders in the populations of both surveyed contaminated and surveyed control settlements studied under the [IAEA] Project, but no health disorders that could be attributed directly to radiation exposure. The accident had substantial negative psychological consequences in terms of anxiety and stress due to the continuing and high levels of uncertainty, the occurrence of which extended beyond the contaminated areas of concern. These were compounded by socioeconomic and political changes occurring in the USSR.*

> *The official data that were examined did not indicate a marked increase in the incidence of leukaemia or cancers. However, the data were not detailed enough to exclude the possibility of an increase in the incidence of certain tumour types. Reported absorbed thyroid dose estimates in children are such that there may be a statistically detectable increase in the incidence of thyroid tumours in future.*

> *On the basis of the doses estimated by the project and currently accepted radiation risks estimates, future increases over the natural incidence of cancers or hereditary effects would be difficult to discern, even with large and well designed long-term epidemiological studies.* (p. 32)

The IAEA study has attracted considerable criticism. For example, it has been pointed out (*The Chernobyl Legacy*, Friends of the Earth briefing sheet, 1991) that the radiation doses and health effects on the workers and soldiers involved in coping with the accident are not included in these studies: they say that the number of people exposed in this way was 600 000, and quote rumours that 5 000 of them have already died. However, one of their main concerns is the fact that the IAEA is the United Nations organization responsible for *promoting* the peaceful uses of atomic energy. Indeed they say:

> *the objectives of the IAEA are incompatible with investigating the health consequences of the world's worst nuclear accident. (The Chernobyl Legacy, p. 5)*

Whether or not any adverse health effects in the general population will be *discernible* above their normal incidence may be open to question. However, it is not disputed that there will be some, although estimates of how many differ widely. An early estimate of the number of excess cancer deaths (*Chernobyl*, CEGB, October 1986, p. 18) was that in the 40 years following the accident there would be between 8 000 and 34 000 excess cancer deaths in the western USSR, and 2 000 in Europe, of which 40 would be in the UK—a total of between 10 000 and 36 000.

The IAEA-estimated collective dose for the three republics most affected is 50 000 man Sv for the 74 years until 2060 (IAEA report cited above, p. 41). Assuming the figure of one excess death per 20 man Sv that we worked out in Section 5.3, this would give a figure of 2 500 deaths in the population of the three republics. Estimates by individuals quoted by Greenpeace (*Questions and Answers on Nuclear Energy*, 1989, p. 30) are factors of ten higher than the official estimates, and go up to half a million deaths world wide.

How is it possible to get such widely varying estimates? To make *any* estimate of the health effects requires knowledge of the initial distribution of the radioisotopes, and of the population. It is then necessary, for example, to model the way in which the radioisotopes are absorbed by the soil, both to estimate the external radiation dose to the population and to see how the radioisotopes are incorporated into the food chain. To estimate the effects of contaminated food requires knowledge of the dietary habits of the affected populations. Many of these factors are poorly known, which accounts for some of the diversity in the estimates made. Then there are differences of opinion on the stochastic effects of radiation, which were discussed in Chapter 4. Finally, individuals or groups wishing to reinforce a point of view would be very unusual indeed if they did not make assumptions that favoured their case. Given all these factors, perhaps it is not surprising that the estimates made are so diverse.

6.4.6 The implications of the accident for reactor operation

At a meeting held by the IAEA in August 1986, the delegation from the USSR presented their views regarding the lessons to be learned from the accident. In the report of this meeting (USSR State Committee on the Utilization of Atomic Energy, *The Accident at Chernobyl and its Consequences*, 1986) they said:

> *In general, the nuclear safety standards in force do not require revision. However, more careful verification of their implementation in practice is necessary. The quality of training and retraining of staff needs to be improved... (Part I, p. 42)*

The parallel with the comments made following the Three Mile Island accident is striking.

Activity 6.2

Having seen how human error contributed to both the Three Mile Island and the Chernobyl accidents, have your views on nuclear power changed in any way? Write two or three sentences to explain why they have or have not changed.

6.4.7 An interlude: points of view in the press

The nature of the Chernobyl accident, and the way in which the picture of what happened had to be patiently pieced together over the weeks that followed it, meant that the press coverage was extensive and often dramatic. Even the most considered articles contained speculation ('unconfirmed reports') that was subsequently proved to be wrong. One suspects that some papers even *made up* experts and their views. For example, the *Sun* carried a story (Figure 6.10) suggesting that the Chernobyl reactor would, in four million years, go through the Earth and finish up in New Zealand. The fact that part of the centre of the Earth is molten, and makes this impossible, seems to have escaped their 'expert'! The hypothesis also ignores the influence of gravity, which attracts matter to the centre of the Earth, not the opposite side of the planet.

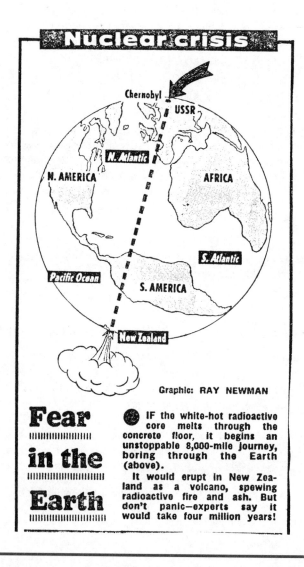

Figure 6.10   From the *Sun*, 10 May 1986.

Not surprisingly, the nuclear industry reacted quickly and strongly to any suggestion that such an accident could happen in the UK, and articles and briefings like those in Extract 6.1 quickly appeared.

Extract 6.1 From *The People*, 11 May 1986.

Britain's power chief answers your fears over Chernobyl

IT COULDN'T HAPPEN HERE

says Lord Marshall, Chairman of the Central Electricity Generating Board

Safety in depth: that sums up Britain's nuclear power safety policy.

It is a safeguard policy that begins long before a nuclear station is built.

And it continues through the design, construction and operation of the station until the end of its life.

Under this policy, each of our nuclear power stations is built to high standards, with immensely strong containment of the nuclear reactor, and operated by highly-trained staff. About one in ten of all power station staff is employed on safety.

The layer of safety

Each station is equipped with protective instruments and controls duplicated and triplicated where necessary to ensure that it can be shut down safely in the event of any fault.

Furthermore, in everything we do we assume an extreme form of Murphy's law. If anything *can* go wrong, we assume it *will* go wrong and we provide a layer of safety to guard against that.

And if that layer of safety is actually called upon, we assume that will go wrong also and provide yet another layer of safety to make sure that the operators and the public are properly protected. That is what we mean by safety in depth.

Never fully satisfied

Safety policy is also watched over by independent nuclear inspectors who act like "nagging wives". They are never fully satisfied with safety standards.

These measures throw a chain-of-safety around our nuclear plants that has been demonstrated by nearly 25 years of safe operation in the UK.

This comprehensive approach to safety and the CEGB's outstanding record gives great confidence in the security of our stations. It also provides the basis for my firm conviction that a Chernobyl-type of accident could not happen in the UK.

But there is another fundamental reason to support this view. The Chernobyl reactor was a design unique to Russia. It would never have been built or licensed in Britain.

The UK nuclear industry looked at the Russian system ten years ago, and rejected it—because against British safety standards it suffered from a number of serious defects. In other words, it never got to first base.

We also looked in great detail at another type of reactor with some similar characteristics to the Chernobyl plant.

However, while this reactor had a number of major advantages over the Russian system it, too, was turned down—because a full-scale version couldn't be made to meet the UK's high safety standards economically.

Prosperity and jobs

Nuclear power helps us to get the price of electricity down. During the miners' strike it demonstrated its importance by helping to keep the lights on.

Other countries are not going to turn their back on nuclear power. The French, for example, will soon have 56 reactors operating giving them the advantage of very cheap electricity.

If we abandon nuclear power, our whole manufacturing industry would be put at a great disadvantage in competing in world markets.

This would seriously affect the country's prosperity and employment prospects.

We have every reason to be very proud of our nuclear safety record and our approach to nuclear safety. It is the best in the world.

It would be tragic if an accident in a Russian plant, which bears no resemblance to anything we use here, persuaded us to forego our modest plans for the future development of nuclear power—and the benefits this would bring.

Safety measures
type of reactor

Activity 6.3

Extract 6.1 gives the views of the (then) Chairman of the CEGB on the Chernobyl accident.

(a) Why did he claim that a Chernobyl-type accident could not happen in the UK?

No

(b) Is enough information given to allow you to compare working practices in the UK and the former USSR, and hence to judge the validity of the claims made?

(c) Does the article appear authoritative and, if so, how does it achieve this?

Partly, by giving quotes as 'fact'
without room for hesitation'.

Equally quickly, those opposed to nuclear power sought to take advantage of the situation.

Activity 6.4

Extract 6.2 is from an advertisement placed by Friends of the Earth.

No

(a) Does the advertisement say what the actual risk from the Chernobyl accident is?

Implies there are none

(b) What does it imply about the risks of other energy sources, which it quotes as being alternatives?

(c) Is the advertisement effective and, if so, how does it achieve this?
Possibly voiced, uses human cost as its crutch, points out other figures

Extract 6.2   From the *Daily Telegraph*, 8 May 1986.

A positive message on Chernobyl from Friends of the Earth

IT NEED NOT HAPPEN AGAIN

The nuclear accident happened 2 000 miles away. Yet we in Britain were still warned not to drink fresh rainwater.

What does that say about the likely effects of a similar accident here? And does anyone now doubt that it could happen?

Of course we are tempted to say "We told you so", for we warned at the Windscale Inquiry, we warned at the Sizewell Inquiry, we have warned for fifteen years that there was danger in the combination of human fallibility and technology with such unprecedented capacity for environmental harm.

But there's no satisfaction in being proved right on this issue. What matters is that a similar disaster in Britain is as avoidable as it is at present predictable.

Nuclear power amounts to only four per cent of Britain's energy supply.

Think about it.

… all that danger, all that cost, all the unsolved problems of waste for *just four per cent* of our energy.

When we have 300 years' supply of coal.

When we have North Sea Oil.

When we have North Sea Gas.

When we haven't even started to conserve energy.

When we haven't even started to explore alternative energy sources … such as sun, wind and waves.

You now know that the human cost of nuclear energy is too high. This is the time to add your voice to ours. Demand that no more nuclear power stations are built.

So, join us now. Or at least give us financial support. Someone has to speak on your behalf—we have the expertise and experience to do it.

Both the nuclear industry and those opposed to nuclear power made statements before they had any detailed knowledge of either the cause, progress, or possible outcome of the accident. However, even quite soon after the accident, some writers managed to have a perspective on it, and on the general lessons that might be learnt from it.

Activity 6.5

(a) What lessons did the writer of Extract 6.3 think could be learnt from the Chernobyl accident?

Accident could happen anywhere. Catas. of nuc. war Technology

(b) What, if any, contribution do his views make to a consideration of the issues arising from the Chernobyl accident?

humane viewpoint, not trying to score point

Extract 6.3   From the *Sunday Mirror*, 11 May 1986.

Chernobyl

TIME FOR THE GLOATING TO STOP

by Charles Wilberforce

The explosion at Chernobyl was a disaster for the Soviet Union, for its technology and for the face it presents to the world.

Unfortunately, some people in the West have seen it as a reason for gloating. That is the last thing it is.

For a start, it could happen here, or if not here in the United States, or in France, or in West Germany.

It isn't very likely. But it is not impossible and we shouldn't forget it.

In one way, Chernobyl has helped the world, because it has demonstrated what a limitless catastrophe a nuclear war would be.

If one damaged power station reactor can panic and poison—however slightly—a continent, imagine what hydrogen bombs hurled by intercontinental rockets would do!

Of course, the Russians were unforgivably slow in warning their neighbours about the dangers of radioactive fallout from Chernobyl.

But there is some evidence now that the Russian leadership was not immediately told of the gravity of the accident. Whatever the origins of the delay, it is the consequences we have to deal with.

The industrial West—the U.S., Britain, West Germany and France, in particular—have the knowledge and the technology to help Russians to contain the effects of the explosion. We must give it freely.

Then we should all—East and West—get together to see what can be done to reduce the possibility of Chernobyl happening again.

And if it should, how the expertise of all of our scientists, engineers and physicians can be used instantly to minimise the damage.

That—not scoring propaganda points—would be a true victory for mankind.

The concerns in Britain centred on the contamination of locally produced foodstuffs. There were scares that resulted in a drop in the consumption of milk and vegetables. Although these scares were short lived, and turned out to be unjustified, there was a general lack of information on which people could make judgements, which heightened the sense of crisis. Concern over the radioactive caesium content of sheep from some areas of North Wales and the Lake District, however, turned out to have been justified. Heavy rainfall over these two regions at the time the radioactive cloud from Chernobyl was overhead meant that levels of ^{137}Cs contamination were comparatively high (we gave the figure earlier as 20–25 kBq m^{-2}). Because of the unusual ecology of these upland areas, the caesium did not become bound into the soil, but was recycled between the sheep and the grass for years after the accident. As a result, meat from many sheep in these areas was deemed to be unfit for human consumption for some time.

Summary of Section 6.4

1   In the version of the RBMK reactor used at Chernobyl there was a positive void coefficient, which was especially marked at low power.

2   In the 1986 accident at Chernobyl, xenon poisoning led the operators to run the reactor at low power with nearly all the control rods withdrawn from the core. To do this, they blocked automatic emergency shutdown mechanisms.

3   The positive void coefficient then led to an uncontrollable power surge, which blew the cap off the reactor; sections of the core disintegrated and melted, the graphite caught fire, and a radioactive cloud was dispersed over much of Europe.

4   It is difficult to estimate the excess cancer deaths that the accident will cause. World-wide figures varying from tens of thousands to hundreds of thousands have been proposed by different organizations.

From the *Daily Mail*, 24 June 1986.

'I don't care what the Ministry of Agriculture says, the sheep haven't been the same since Chernobyl . . .'

6.5 Risk assessment for reactor accidents

Underlying the cost of your car insurance is an assessment of the risk that you will have a car accident. The assessment draws on a large bank of statistics about accidents that have happened in the past, and which are considered to be like the accidents that you may have. The problem with making similar assessments of the risk of an accident to a nuclear reactor is that accidents like the three we have described in Sections 6.2–6.4 are rare. There have been numerous *incidents* involving nuclear reactors, many of which are chronicled in *The Greenpeace Book of the Nuclear Age* (1989), but the three we have described are the principal ones that have given rise to an *actual* risk to the public. Hence we have very little experience from which quantitative estimates of risk can be determined.

In order to evaluate the safety of nuclear power, it is therefore necessary to try to *predict* what accidents could occur, and what their outcome might be—not just the spectacular, 'worst case' ones, but *all* types of accident. If this study can be combined with an estimate of the *probability* that the accidents concerned *will* happen, then the basis for making an informed judgement is available. This is clearly a difficult task; it is, however, a necessary one. In the next section we describe a technique used in tackling it.

6.5.1 Probabilistic risk analysis

In activities as complex as nuclear power—as indeed for many commercial and industrial activities—there are a whole host of events that can give rise to risks. One method of studying these events and establishing the risks and the probabilities of them occurring is **probabilistic risk analysis**.

Event trees and probabilistic risk analysis

Suppose that a pump in a nuclear reactor breaks down, and that a safety valve that should open in response to this breakdown fails to do so. If the probability of the pump failing during its agreed lifetime is 1 in 1 000, and that of valve failure is 1 in 100, then, assuming the two events to be independent, the probability that the sequence pump failure–valve failure will occur is obtained by multiplying these two probabilities together; it is 1 in 100 000. The other possible sequences in this situation are shown in the **event tree** in Figure 6.11.

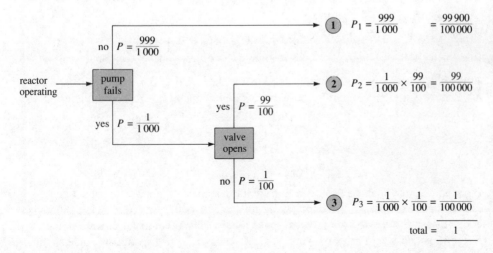

Figure 6.11 An event tree for a sequence of two events in a nuclear reactor in which a safety valve is supposed to open in response to a pump failure. If the probabilities, P, of pump *failure* and of valve *failure* are 1/1 000 and 1/100, respectively, then those of the pump and valve *functioning* must be 999/1 000 and 99/100, respectively, because failure and functioning together cover all possible events. Hence their probabilities must add up to 1. The probability of two independent sequential events is obtained by multiplying their probabilities together. Thus, the sequence pump fails–valve opens has a probability of 1/1 000 × 99/100. There are three possible outcomes whose probabilities, P_1, P_2 and P_3, add up to 1, because they cover all contingencies arising from the combination of events. This is most easily shown by converting the probabilities of the outcomes to fractions with the same denominator, as in the right-hand column.

To check that you understand the principles underlying Figure 6.11, you should now do Activity 6.6.

Activity 6.6   *You should spend up to 15 minutes on this activity.*

In Figure 6.11, two possible sequential failures can interrupt the operation of the reactor: pump failure followed by valve failure. In this activity you will deal with a possible sequence of three events, which is most easily described by stating one of the outcomes: a pump fails, a meter signals the failure, and an operator acts on the signal. The individual probabilities of pump failure, meter failure and operator failure are 1/100, 1/100 and 1/20, respectively.

1   Construct an event tree like Figure 6.11 showing the possible sequences of events.

2   How many outcomes are there, and what is the probability of each?

3   What is the sum of the probabilities of the outcomes, and why does it have this value?

4   If an accident sequence is one in which the last event is undesirable, which of the outcomes terminate in an accident sequence?

Although the exercise in Activity 6.6 was very like the one embodied in Figure 6.11, it did, in one sense, include a new feature: the possibility of operator error drew attention to the human element. In risk analysis, the human element is extremely important, and this is why the prospect of replacing people by computers (Box 6.1) is in many ways attractive.

So far, we have considered sequences of just two or three events whose outcome is considered dangerous. In a full-blooded probabilistic risk analysis of a nuclear reactor, it is necessary to write down all conceivable event trees in order to find those that terminate in accidents, and then, by estimating the probability of each step, the probability of the accidents can be calculated. Clearly this is not easy. Where operators appear in the trees, the probability of human errors must be estimated, and this is notoriously difficult. The assessment of the risks of *mechanical* failure is in some instances easier. The probability of failure of standard pumps or valves, for example, will be available from statistical data about failures in other industries. However, in other cases, such as reactor pressure vessels, there are no precedents, and the probability of failure has to be estimated.

Thus, the value of probabilistic risk assessment depends on three factors:

(i)   how well the probabilities of failure are known for all the mechanical and electrical components featuring in the event trees;

(ii)   whether the human factor has been properly allowed for;

(iii)   whether all possible event trees leading to an accident have been analysed.

In the end, the answers are only as good as the data used to derive them; if the data are faulty, then the answers will be wrong. We now consider a real example of an application of probabilistic risk analysis to nuclear reactors.

6.6 The Rasmussen Report

Probabilistic risk analysis has been fairly widely applied to nuclear reactors and featured, for example, in the safety study for the Sizewell-B PWR. However, the most comprehensive study to date (at least, publicly available) remains the 'Rasmussen Report', named after Professor Norman C. Rasmussen, who was director of the study. The report was entitled *Reactor Safety Study: an Assessment of Accident Risks in US Commercial Nuclear Power Plants*, and was published in 1975 by the US Nuclear Regulatory Commission.

Risk assessment is a time-consuming activity: over a three-year period, 60 people put in 70 person years of work. What they did was to examine the risks that could arise from the 100 nuclear reactors distributed across the USA, this being the number that were expected to be running in 1981.

First, they constructed thousands of event trees, involving about 130 000 potential accident sequences, including those that derive from human error, to find out which ones led to the release of radioactivity into the environment. (Remember: every event tree has many branches, and hence potential accident sequences; the event tree in Activity 6.6 had two potential accident sequences in it.) They did this both for the reactors and for the spent fuel stores, which are a potential risk by virtue of the radioactivity they contain.

Next, they investigated the probabilities of the event sequences leading to a release of radioactivity. A total of 140 000 different possible combinations of radioactive release, weather pattern and population distribution around six 'typical' sites were then analysed in order to build up a statistical picture of where any released radioactivity would go.

Using these data, the radiation doses that the population would receive were calculated, taking into account evacuation procedures, and these were used to find out how many people would die, either immediately or later (from diseases that take some time to become manifest), or become ill (for example from thyroid cancer, caused by the uptake of radioactive iodine). They also considered how many cases of genetic damage there would be. Finally, they looked at the land areas which would become contaminated, and at the total cost of the accidents. Figure 6.12 shows a summary of these steps.

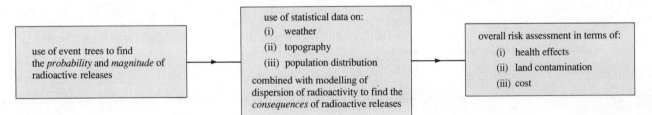

Figure 6.12 The steps used in making probabilistic risk analysis of nuclear reactors.

Of course, accidents to reactors can arise from external causes, and the possibility that earthquakes, tornadoes, floods, crashing aircraft and steam turbine failures could start an accident sequence were also investigated. The only causes that were not tackled were sabotage and acts of war, although the degree to which reactors would be able to withstand such attacks was discussed.

The whole report runs to 12 volumes, and the Executive Summary alone is 198 pages long! There are a host of conclusions relating to all the possible effects we have discussed, but the overall conclusion was that the annual average risk of death for an individual in the USA from accidents in the 100 reactors was far lower than from other accidents arising from human activities—for example air (plane) crashes, fires and dam failures—and also far less than from most natural events—for example hurricanes and earthquakes. In fact only falling meteorites posed a comparably low danger.

Figures 6.13 and 6.14 summarize these conclusions. These figures show the number of times per year that a particular phenomenon will cause more than a certain number of deaths in the USA. A lower limit for the number of deaths caused by a particular type of disaster is plotted along the *x* axis; along the *y* axis is plotted the frequency, or number of occasions per year, with which that type of disaster causes more than this limiting number of deaths. Let us consider an example to show how these figures can be used.

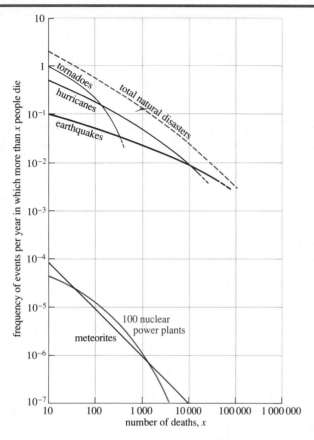

Figure 6.13 Frequency with which more than a specified number of people die in a particular natural disaster in the USA compared with 100 US nuclear power plants.

Figure 6.14 Frequency with which more than a specified number of people die in disasters in the USA consequent upon human activities.

▷ From Figure 6.13, what is the frequency with which there will be an earthquake resulting in the death of more than 10 000 people?

▶ Looking at the curve for earthquakes in Figure 6.13, you will see that the frequency corresponding to 10 000 deaths is about 10^{-2} events per year. 10^{-2} is 0.01 or 1/100: in accordance with the discussion of Box 4.3, this means that there is a 1 in 100 chance per year of an earthquake killing more than 10 000 people. Another way of looking at this is that on average there is likely to be an earthquake like this once every 100 years.

Question 6.1 Use Figures 6.13 and 6.14 to answer the following questions:

(a) What is the risk per year of an event in which at least 1 000 people will die due to accidents involving the 100 US nuclear power plants?

(b) By how much would the risks from the 100 nuclear power plants have to be increased so that the risk per year of at least 100 people dying from a reactor accident becomes the same as that from air crashes?

(c) By how much is the risk of 1 000 or more people dying from any one of the natural disasters considered in Figure 6.13 greater than the risks of them dying from an accident in one of the 100 nuclear power plants?

(Note that the graphs have logarithmic scales on both axes. When answering this question, it is sufficient to read off the points on the frequency axis to the nearest power of 10.)

Since most people feel, probably quite rightly, that the risk of death from a falling meteorite is very remote, and also feel that the risk from nuclear reactors must be much more, this comparison often provokes the feeling that the results of the study must be wrong. But before reaching such a conclusion, let us first see what limitations were *expected* when the study was commissioned. In a statement of objectives, the US Atomic Commission stated:

...although the results of this study of necessity will be imprecise in some aspects, the study nevertheless will provide an important first step in the development of quantitative risk analysis methods. (Executive Summary, p. 2)

The authors of the report assigned an *order of magnitude* of uncertainty to their results; that is, their estimate of the risks could be wrong by a *factor of ten up or down*. Looking back at Figures 6.13 and 6.14, increasing the risks by a factor of ten still does not make the risks from nuclear accidents more serious than the other causes cited.

Another criticism is that they may have overlooked some vitally important event sequences. It must be said immediately that *this can never be ruled out*. However, consider what the report itself says about this possibility:

It is important to understand that the Reactor Safety Study does not purport to have included in its results contributions from all conceivable accidents ... The important question is not whether all *contributions have been included, but whether* significant *contributions to risk have been included ... The goal of an analysis is to include a sufficient number of significant contributions so that the results are insensitive to further contributions.* (Executive Summary, p. 147)

Irrespective of whether the *absolute* levels of risk are accurately predicted using event-tree analysis, the methodology is undoubtedly valuable in identifying the *most important* risks. It *should* therefore allow the reactor designers to concentrate on reducing these risks, in order to minimize the overall risks.

One of the comments made after the Three Mile Island accident was that reactor designers had not used the Rasmussen Report in this way. For example, the report showed that, in PWRs, the risk that the core would melt if a pressure release valve failed to close following a sudden change in reactor operation was important in 5 of the 12 event sequences which led to the core melting; the overall probability per year per reactor of the core melting following such sudden changes in reactor operation was 1.6×10^{-5}, of which these 5 sequences contributed 0.4×10^{-5} (that is, 25%). Yet, despite this, and despite the fact that there had been similar failures elsewhere, the operators at Three Mile Island were not even alerted to the problem. This suggests that studies like the Rasmussen Report might be of considerable practical benefit to reactor designers and operators in the future.

Summary of Sections 6.5 and 6.6

1   Because nuclear accidents are so infrequent, the risk that one will occur cannot be estimated from experience or statistics.

2   Risk estimates can be made by specifying all conceivable sequences of events which culminate in accidents, and then calculating the risks of the accidents as multiples of the estimated probabilities of the sequential contributing events. The probabilities of at least some of the contributing events *will* be derivable from experience and statistics.

3   The best-known attempt to do this is the Rasmussen Report, which assessed the risks associated with 100 commercial nuclear power plants operating in the USA.

4   The Rasmussen Report concluded that the risks of death from the 100 nuclear power plants were comparable with those from meteorites, and hence very low. The risk estimates in the report were assigned uncertainties of a factor of ten, up or down.

7 The economics of nuclear power

In this chapter we shall be considering the problems of estimating the cost of generating electricity using nuclear power, and of making comparisons with the cost of other power generating systems, notably coal. However, in the past the choice of a generating system has been influenced not just by costs, but by the social and political attitudes that prevailed when the choice was made. Because we want you to bear this in mind, we begin by surveying some of the changes that have taken place since the end of the Second World War. Only then shall we turn to the question of economics.

7.1 Energy and politics

After the Second World War, in 1945, the UK government sought to restore the national economy by assuming responsibility for a wide range of industrial activities—for example, running the railways, coal mines, airlines and steel industry, and also for the supply of electricity, gas and water. In common with other countries, it also assumed responsibility for developments relating to nuclear energy, and, in due course, for the introduction of nuclear power stations.

It was then considered to be in the public interest for these and other activities (for example farming) to receive government funds, either to support development programmes or simply to maintain the activity. Thus, many industries were subsidized, either directly or through favourable loans, and their long-term development and responsibilities were considered to be the concern of central government. This was particularly true in the energy industries.

During the period we are speaking of, the British government's energy investment decisions were partly motivated by a sense that the UK's energy supplies were vulnerable, and would benefit from diversification. In 1950, 87% of the UK's total energy supply came from coal, compared to 30% in 1990. This made governments feel vulnerable to action by the miners, a fear later substantiated by the bitter miners' strikes of 1972 and 1974. In both of these, the miners won large wage increases, and in the second they drove the government to call a general election, which it lost. It must be remembered that, before North Sea gas ('natural gas') was discovered and started to be used in 1968, gas was produced from coal or oil, so that gas supplies were also vulnerable if supplies of these primary fuels were disrupted. The perceived vulnerability of coal to industrial action led to an increased use of Middle Eastern oil for electricity production in the late 1960s and early 1970s. North Sea oil was not then available, and events such as the closure of the Suez Canal in 1956 gave rise to considerable concern over the security of oil supplies. Nevertheless, in the UK the use of oil for electricity production expanded until the severe price rises imposed by the oil-producing countries in 1973–4.

Because of the national importance of electricity supplies, the use of nuclear power for electricity production was seen by successive governments in the 1960s and 1970s as a way of reducing the dependence on coal and, to a lesser extent, oil. In *The*

Power of the State: Economic Questions over Nuclear Generation (Adam Smith Institute, 1991) Professor Colin Robinson set out the basis of this strategy very clearly:

> *Diversification of supply sources is clearly one appropriate means of enhancing security. Labour disputes, technical difficulties, monopolistic action by suppliers, natural catastrophes and other events which affect one source are unlikely to affect others at the same time; thus they become manageable events. There is a case for a judicious mix of supply sources which, in the electricity supply industry means a mixture of primary fuels (both indigenous and imported), of generators (home and foreign), of technologies and of scales of operation. Of course, what constitutes a 'judicious mix' is largely a matter of judgement.* (p. 50)

If now, in the 1990s, the response to fears in the 1950s and 1960s is difficult to understand, this is because of changed circumstances and changed attitudes in the UK. For example, the state has surrendered control of most energy industries to the private sector, and the primary concern of these industries is now to remain solvent and make profits. The sense of vulnerability to domestic coalminers' strikes has faded following the failure of the strike of 1984–5 and the increased reliance on cheaper imported coal. Anxiety over reliability of supply has been replaced by other worries. Today, for example, there is concern that the increase in the atmospheric CO_2 concentration brought about, in the main, by burning fossil fuels could bring about significant changes in the world's climate. But it is conceivable that, in 50 or 100 years time, these fears will be shown to be misplaced, or that other ones may have assumed a greater priority.

7.1.1 Electricity: problems of demand and supply

The use of electricity is central to most aspects of modern living. We take its supply for granted and we expect the supply to be sufficient to meet all our varying demands; you do not have to give notice to the electricity company when you want to switch on your washing machine. This means that the electricity supply has to be very *flexible*, to cope with changes in demand. This change in demand has two features. Firstly, there are *seasonal* changes (Figure 7.1); for example, in the UK we use more electricity in the winter than in the summer. Secondly, there are *daily* changes (Figure 7.2); we use less electricity at night than during the day. There are also peaks in the daily demand associated with, for example, meal times and with the ending of popular TV programmes, when everyone rushes to make a cup of tea.

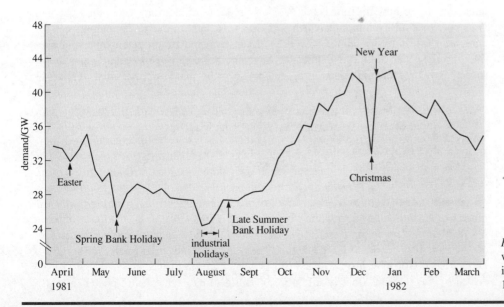

Figure 7.1 The variation in the weekly demand for electricity during the year April 1981–March 1982.

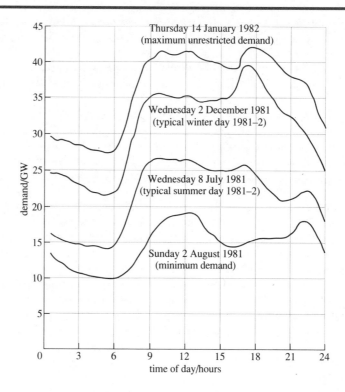

Figure 7.2 The variation in the hourly demand for electricity during four days at different times of 1981–2.

Question 7.1 Using Figures 7.1 and 7.2, what are the minimum and maximum electricity demands over (a) the whole year, and (b) Wednesday, 2 December 1981?

What does having to meet such varying demands mean to companies like Powergen and National Power, who, since the 1990 privatization, supply our electricity? It means that:

(i) they must have enough electricity generating plant to meet the *maximum demand* at any one time;

(ii) they must be able to meet quite rapid *changes* in demand.

The minimum amount of electricity which has to be supplied is called the **base load**; the power stations producing this are run steadily without stopping. The daily and seasonal changes above this base load have to be met by using extra generating systems, which are brought in and out of use as required. Electricity generating plant that does not work all the time, or which works at less than its maximum output, will generate less electricity than if it were working at its maximum potential output all the time. The **load factor** is defined as the average amount of electricity generated (usually over a year) expressed as a percentage of the maximum potential output.

▷ A power station with a maximum potential output of 850 MW produces an average output of 570 MW over a year. What is its load factor?

▶ The load factor averaged over a year for this station is (570 MW/850 MW) × 100%, that is, 67%.

The power output from most electricity generators is flexible; they can cope with a range of demands up to a maximum value. If the demand exceeds this maximum, then it will be necessary to start up another generator, but how quickly another generator can be started depends on the generator. The pumped storage, hydroelectric system at Dinorwig, in North Wales can reach full power output in 10 seconds; generators driven by gas turbines (that is, aircraft 'jet' engines) can achieve full power in less than 5 minutes, and coal- or oil-fired stations need from 1 to 8 hours. Finally, nuclear power stations take between 12 hours and 2 days when starting after a complete shut-down.

Renewable energy sources such as wind, wave or tidal power come in a different category, in that their maximum output varies with time. The output of wind- or wave-driven generators depends on the weather, which is unpredictable; tidal power output is predictable but linked to tidal ebb and flow.

▷ Since the ability of wind, wave and tidal power to respond to demand is limited, what are the implications of using them for electricity generation?

▸ Sufficient extra generating plant would have to be available to cover periods when the output from such 'alternative' energy sources is low. Providing this additional generating plant would increase the overall cost of using renewable sources of energy.

Their variability, or unpredictability, of supply is one of the disadvantages of using renewable sources of energy. This is because, at present, electricity is chiefly supplied *on demand*, not when it is *available*.

Summary of Section 7.1

1 Since the 1970s, there has been a trend towards diminished state control of energy production in the UK. The energy industries have therefore been judged more in terms of cost and short-term returns on investment, rather than by other criteria such as security of supply and balanced diversity, which were favourable to nuclear power, and supported by central planners.

2 The seasonal and daily variations in demand for electricity require a flexible means of supply. There must be some power stations that can work continuously to provide the base load. Additions to the base load are managed by starting up or shutting down other generators when necessary.

7.2 Estimating the costs of future projects

The problems involved in assessing the economic viability of a nuclear power station are no different in principle from those of any other large-scale, capital-intensive, long-term industrial project, for example building the Channel tunnel or developing a new aeroplane.

The first problem is that the *planning and building* times are long (5–15 years or more) and, once built, the projects are in operation for upwards of 20 years. Hence, whether it is judged that there is a *need* for a power station, an aeroplane or a tunnel, and whether the investment in it can be justified, depends in part on what is likely to be happening in 30–50 years' time. This, of course, is very hard to predict.

The second problem is that estimating the **capital cost** of a large project—that is, the money that has to be spent to build it—is difficult, particularly if there is little or no previous experience of that type of construction. Additionally, changes may be required while construction is in progress, because of, for example, improvements in the safety standards required. Hence, initial estimates of costs are nearly always wrong and are inevitably underestimates; indeed, underestimates of a factor of 2 or more are not uncommon. Such 'cost over-runs', as they are called, often only become apparent when the project is too far advanced to call a halt. The investors then have to supply extra funds, or risk losing everything that they have invested.

A third problem is that the money borrowed to build the power station has to be repaid, together with the *interest* on it. The longer the period over which repayment is made, the greater the total interest that has to be paid, but the lower the *annual* repayments; if you have ever negotiated a mortgage for a house, you will recognize this pattern.

The cost of repaying a loan made to build a power station, together with the interest on it, forms part of the cost of the electricity provided to the consumer. Therefore, the shorter the repayment period, the higher the cost of electricity from the power station while the loan is being repaid. However, if interest rates change during the project, then, again, financial estimates will be seriously in error.

A fourth problem that arises when construction times are long is that contractors cannot wait to be paid until the end of the project; hence they receive payments as the project proceeds. This money will normally be borrowed, and interest will have to be paid on it while construction is in progress—that is, *before* there is any income from the project. This is called *interest during construction*. If the construction takes longer than planned, the interest paid during construction will therefore be increased.

Finally, once the project is operational, unforeseen problems may arise. For example, the **running costs** for the plant, or the demand for the product, may change. Alternative, more attractive ways of doing the same thing may be developed, or additional investment may be needed to meet new safety requirements.

One example of an important project that has been influenced by these factors is the Channel tunnel. Let us consider it more closely.

Activity 7.1

The original estimated cost of the Channel tunnel was £4 700 million (£4.7 billion). Now read Extract 7.1.

(a) By what factor does the new estimated final cost exceed the original one?

(b) What reasons are given for the latest increases in cost?

(c) Why could estimates of income in 1993 be wrong?

Extract 7.1 From *The Guardian*, 8 October 1991.

Eurotunnel cost rises to £8 billion

Daniel John
Transport correspondent

The cost of building the Channel tunnel has risen to more than £8 billion, because of delays and big safety modifications to the passenger and freight wagons, Eurotunnel disclosed yesterday.

The company must now delay payment of its first dividend to shareholders by a further 12 months, to the year 2000, because of the extra costs and a forecast drop in revenue due to a limited launch of services.

The announcement wiped £150 million off Eurotunnel's £2.6 billion stock market value after shares fell 28p to 496p.

The company is also facing a huge bill for extra costs which the builders, Transmanche Link, claim have arisen during the fitting out of the tunnel.

TML, a consortium of French and British companies, says it is entitled to additional payments of £800 million at 1985 prices, which would take the final bill for kitting out to £1.27 billion.

Eurotunnel has contested the claim, and has only budgeted £200 million for the alleged costs. An independent arbitrator has been asked to resolve the dispute.

TML says installing equipment in parts of the completed tunnels is up to six months behind schedule. It is asking Eurotunnel to pay to bring the work back to its original timetable.

Eurotunnel claims the work can be finished in time for the transport system to be commissioned at the end of next year. Sir Alistair Morton, chief executive of Eurotunnel, said TML's claim had not been substantiated.

"Months of valuable time have been lost, though they can be recovered with an effort," he said yesterday, after unveiling the company's half year results.

Eurotunnel also disclosed that fewer people and less freight than planned will be able to use the tunnel after it opens in June 1993.

As *The Guardian* revealed on Saturday, the company's traffic forecasts have had to be cut because of delays over design changes to the shuttle trains which will carry cars and lorries between the terminals at Folkestone in Kent and Coquelles near Calais.

The commission responsible for overseeing safety ordered doors to be widened on the passenger shuttles and the semi-open lorry wagons to be enclosed. Eurotunnel expects to operate a limited service for the first three months after June, but should be fully operating by autumn.

The delays and the redesign will add a further £440 million to £7.61 billion which Eurotunnel estimated last year it would need to finish the project and run the system. The phased opening is likely to cut by £147 million the amount it expects to earn in the first year.

Summary of Section 7.2

1   The financial success of *any* commercial project depends on accurately estimating the *capital cost* of the project, its *running costs* and the *demand* for the product.

2   Factors outside the control of the project management which can change these estimates significantly include changes in interest rates, in safety requirements or simply in demand.

3   Finally, in very complex, long-term projects, particularly those for which there is little previous experience, estimating development and construction costs can be very difficult. The estimates are more often than not in error, and costs are invariably underestimated.

7.3 Cash flows in long-term projects

In this section, we shall introduce a method of assessing whether a long-term project like the building of a nuclear power station is economically worth while, and of comparing it in this respect with alternative projects such as the building of fossil-fuelled power plants. We begin with some basic terminology.

The amount of money which 'flows' into a project is the income; the amount that flows out, the expenditure, is known in economic terms as 'costs'. Together, the income and the costs make up the **cash flow**. The net cash flow is the difference between the income and the costs of a project in any period, and varies during the life of the project. During the building of a nuclear power station, money will be spent (capital will be invested), and the cash flow will be *out* of the project; the convention is that cash flow out is negative. When the power station is operating, there will be income from the sale of the electricity, which should exceed expenditure. There will then be (or should be) a net cash flow *into* the project; the cash flow is positive. When the reactor ceases operation, it will have to be decommissioned (that is, taken apart), and the waste stored, and once again there will be an *outwards*, negative, cash flow.

One of the difficulties encountered when trying to judge whether investment in such a project is worth while is that these different cash flows occur at *different times*. They therefore have different *real* values even when they amount to the same sum in, say, £ sterling. Let us be more specific. Suppose in 1970 you lost a £10 note in the upholstery of your armchair. If you were to find it today, you would know that it is worth much less now than it was worth in 1970.

▷ Suggest a reason for this.

▶ One reason is *inflation*: today's £10 note has a lower *real* value; it simply buys fewer goods and services in the shops than a £10 note did in 1970.

Again, let us be specific. Suppose that you carry your lost £10 note to the shops today and you find that, because of inflation, it will buy only what a £1 note would have bought in 1970. This then is one reason why the £10 note has lost value with time.

Those involved in finance, however, recognize a second reason. Suppose that in 1970 you had used the £10 note to buy shares in a computer company that has since proved highly successful in marketing and selling its products. Suppose also that because of this you receive the substantial sum of £200 when you sell those shares today.

▷ Will this £200 buy more or less than the £10 note bought in 1970?

▶ It will buy twice as much: a £10 note today buys what a £1 note did in 1970, so £200 today buys what £20 bought in 1970.

Thus, there are *two* senses in which the value of the £10 note that you mislaid in 1970 has decreased with time: firstly, inflation has eroded its value; secondly, the opportunity has been missed to invest it wisely and make its purchasing power outpace the rate of inflation.

When trying to predict the economic viability of a future long-term project, it is necessary to make corrections for the way in which these two contributions will cause the value of the cash flows to change with time. One of the most widely used techniques is known as *discounted cash flow analysis*. In the form in which we shall use it, the corrections for inflation will have already been made, so the technique will only be making corrections for the second of the two contributions that we have discussed.

7.3.1 Discounted cash flow analysis

The technique of **discounted cash flow analysis** converts the cash flows for different periods into the estimated value they would have had *at the start of the project*. These revised cash flows are called **discounted cash flows** or **present values**. As we implied at the start of this section, some will be positive and some will be negative. If we add them up, the total is called the **net present value**. If the net present value obtained in this way is *positive* then the project would be worth while; if it is negative, the project would not be worth investing in. Let us use an example to clarify the technique and ideas.

Calculating the net present value

We shall consider a cash flow problem for a simplified nuclear power project; it has been simplified in two senses. Firstly, we have given it a short construction time and a short lifetime so that the results can be more readily understood. Secondly, in accordance with the last sentence of Section 7.3, we have already compensated for the effect of inflation, so that we have in effect assumed an inflation rate of zero. This means that throughout the period covered by the project, a £10 note always buys the same quantity of goods at the local supermarket. By smoothing out the effects of inflation in this way, we concentrate our attention on the second, less familiar contribution to the change in value of cash flows with time. These simplifications do not affect the validity of the points we shall make: they only make the arguments easier to follow.

The cash flows for the project are shown in Figure 7.3. Construction of the power station costs £1 000 million, and takes a year. The station then works for 10 years before being closed down, during which time it earns a net annual income of £150 million.

▷ Why is it reasonable that the incomes in years 1–10 should all be identical (£150 million)?

▶ Remember that we have assumed zero inflation; the price of electricity is not changing with time.

After being closed down, the power station is left for 15 years (assumed to be at no cost). Over a period of time the buildings and reactors can be progressively dismantled, any nuclear waste can be disposed of, and eventually the site can be made available for other uses. This process is called **decommissioning**. We shall assume that the costs are covered by the expenditure of £500 million in year 25. All these data are given in columns 1–3 of Table 7.1,

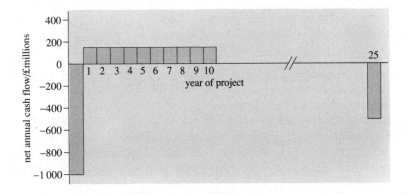

Figure 7.3 Cash flows for a simplified nuclear power station project: in the year of construction there is a cost, or negative cash flow, of −£1 000 million, followed by 10 years of positive cash flow or income of £150 million per year. After 15 years of closure, there is a final decommissioning cost of −£500 million.

Table 7.1 Discounted cash flow analysis for a simplified nuclear power station project.

Year of project	Net annual cash flow		Discounted annual cash flow £ million	
	cost £ million	income £ million	discount rate	
			5%	10%
0	−1 000		−1 000	−1 000
1		150	143	136
2		150	136	124
3		150	130	113
4		150	123	102
5		150	118	93
6		150	112	85
7		150	107	77
8		150	102	70
9		150	97	64
10		150	92	58
25	−500		−148	−46
net present value			12	−124

We must now convert the cash flows for years 1–10 and year 25 given in columns 2 and 3 of Table 7.1 into their equivalent values in year zero. To do this, we ask what sums of money if invested in year zero *in ways other than in the project* would yield the cash flows of columns 2 and 3 at the times specified in column 1? To work this out, we need to specify a future return on investment: an investor's interest rate. At this stage, it is useful to think of it as the interest rate that might be earned by investing in competing projects that are also looking for funding. Once specified, it is called the **discount rate** and, because it is an estimate, it will be subject to considerable uncertainties. For this reason, we shall do the calculation for two values of the discount rate, namely, 5% and 10%.

Suppose we have a sum of money whose present value is P, and we invest it at a specified percentage interest rate, or discount rate, of d. Then in n years time, it will have increased to a sum, F, which is called the **future value**, and is given by

$$F = P \times \left(1 + \frac{d}{100}\right)^n \tag{7.1}$$

Box 7.1 gives the derivation of this relationship because you are unlikely to have come across it before.

7.1

Box 7.1   Derivation of Equation 7.1

The expression is the standard one used to calculate the money that accrues under conditions of compound interest when a sum P, is invested at an annual interest rate of d.

After one year P will have increased to:

$$P + \left(P \times \frac{d}{100}\right) = P \times \left(1 + \frac{d}{100}\right)$$

Remember d is a *percentage*. After two years P will have increased to:

$$P \times \left(1 + \frac{d}{100}\right)\left(1 + \frac{d}{100}\right) \text{ or } P\left(1 + \frac{d}{100}\right)^2$$

After three years it will have increased to:

$$P \times \left(1 + \frac{d}{100}\right)^3$$

and so on. So the power to which $(1 + d/100)$ is raised is numerically equal to the number of years. Hence, after n years, P will have increased to a future value of

$$P \times \left(1 + \frac{d}{100}\right)^n$$

a sum called F in Equation 7.1. ■

A simple rearrangement of Equation 7.1, enables us to calculate the present value, P, from the future value, F:

$$P = \frac{F}{(1 + d/100)^n} \tag{7.2}$$

In columns 4 and 5 of Table 7.1, this equation has been used to calculate the present values (or discounted cash flows) of the figures in columns 2 and 3 at the discount rates of 5% and 10%, respectively. So, for year 1 of the project, Equation 7.2 tells us that the present value of the £150 million net cash flow is £143 million at a discount rate of 5% and £136 million at a discount rate of 10%. In other words, at a 5% discount rate, £143 million invested in other projects in year 0 would be worth £150 million in year 1. Of course, if the discount rate is greater (10%), *less* money (£136 million) invested in year 0 in the other projects will be worth the £150 million in year 1. To be worth £150 million in year 2, even less money need be invested in year 0.

At the bottom of the table, the net present values at the two discount rates have been obtained by adding up the figures in columns 4 and 5, including, in the row before the line, the investment costs needed *now* to cover the costs of decommissioning in year 25. The net present value is £12 million at a discount rate of 5%, which means that the project would be worth while; in other words, it would make more profit than alternative means of investment offering a 5% return. On the other hand, at a discount rate of 10% the net present value is very negative, so that the project would not be worth while: better value for money is obtained by investing in the alternative projects offering a 10% return. You can see from this how important the discount rate can be in determining whether a project is economically viable.

Let us consider why this is so. Notice first that, when they have been discounted, our constant annual cash inflows of £150 million decrease with time: money that will be received in the future is worth less than money received now in the second of the two senses identified in Section 7.3: at any time between 1970 and the present, £10 could have been obtained through the wise investment of less than £10 in 1970. Notice also, that at the higher discount rate of 10%, the decrease in present value is steeper. Thus, the £150 million we expect to obtain in year 10 could then be obtained by investing just £58 million in alternative projects in year zero, and waiting until year 10. The discount rate is the means by which our project is set in competition with others. When the return on alternative investments is higher, our project looks less promising: it is more profitable than the alternatives at a discount rate of 5%, but less profitable at 10%. Nevertheless there is one feature of nuclear power projects which makes a high discount rate an advantage to them.

▷ From Table 7.1, what do you think it is?

▶ Decommissioning: it costs a great deal to wind up this project. A high discount rate means that the future cost of decommissioning has a small present value. Thus, the £500 million needed in year 25 can be covered by investing only £46 million now in other projects at a 10% discount rate.

However, as Table 7.1 shows, this advantage of the high discount rate is not decisive; it is outweighed by the disadvantages of the effects on income and the project is not profitable at 10%.

That concludes our discussion of discounted cash flow analysis. It is essentially a way of predicting the profitability of a new project in comparison with alternative projects that could be invested in. Because it is a prediction, it contains uncertainties. One of the biggest of these arises in specifying the discount rate. We can regard this rate as a guess at the future interest rate that might be earned by investing in the alternatives.

To check your understanding of the principles of discounted cash flow analysis, you should now do Activity 7.2.

Activity 7.2   *You should spend up to 30 minutes on this activity.*

Suppose that having calculated the figures in Table 7.1, the electricity company faces two different sets of objections. The first is from its radiation protection division, who say that delaying the decommissioning will reduce the radiation dose to those involved. The second is from an environmental group, who claim that the cost of decommissioning has been underestimated by a factor of two, and should be £1 000 million. Finally, the company itself plans improvements in station output which will increase the annual cash flow during the 10 years of operation to £165 million.

(a) Calculate the net present values of the project at discount rates of 5% and 10% if Table 7.1 is modified by the single assumption that decommissioning will cost £1 000 million and take place in year 50.

(b) Write out a revised Table 7.1 for the case in which the annual income in the 10 years of operation is £165 million, the discount rate is 10%, and decommissioning, which takes place in year 50, costs £1 000 million. Is the project profitable under these conditions?

7.3.2 Internal rate of return

In Section 7.3.1, we used discounted cash flow methods to analyse profitability in terms of net present values. Alternatively, the analysis can be conducted by using the same method to calculate what is called the **internal rate of return**. This is sometimes also called the real rate of return. The internal rate of return is the discount rate that would yield a net present value of zero. Look at Table 7.1.

▷ Is the internal rate of return for this project closer to 5% or to 10%?

▶ To 5%; the net present value is just above zero at a discount rate of 5%; at 10% it is very negative. We can deduce that the discount rate that makes the net present value zero is between 5 and 10%, but much closer to the lower figure.

In fact, a discount rate of 5.5% sets the net present value of this particular project to zero. This, then, is the internal rate of return in this case.

The internal rate of return required for a project to proceed depends on the industry. One reason for this is that the risks involved vary. In stable industries like food, the expected rate of return is low, because rapid changes in demand are unlikely. In rapidly changing fields like computing, a new product may fail quite unexpectedly, so firms compensate for this by basing their economic judgements on a higher rate of return. Demanding a higher internal rate of return for a project amounts to a demand for increased income and for decreased costs compared with alternative projects with similar risks, in other words a demand that the project should become more competitive.

Thus, as with the discount rate, the value of the internal rate of return defines the competitiveness of the project as an investment, and the higher the rate used the more competitive it has to be. This point is of special importance when we turn to state industries. Here, the government and not the market determines the discount rate or rate of return to be used in project assessments. Whether or not these are comparable with those used in private industry is then a matter of government policy; fixing low values then raises the apparent profitability of the industry concerned, and enables it to reduce the prospective price of its product. Thus, setting a low discount rate or rate of return in this case can be construed as the inclusion of a government subsidy in the assessment. (You will gain a greater appreciation of these points when you read Extract 7.2 at the end of the chapter.)

7.3.3 Net effective cost

Before privatization, a government body, the CEGB, was responsible for electricity production in England and Wales. When making a case for building PWR reactors (of which that at Sizewell, Suffolk, will be the first) they used a method of economic appraisal which allowed them to take into account savings that would result from using a new nuclear power station to generate electricity, rather than continuing to use existing, less efficient, fossil-fuelled ones—in particular, those using coal.

The procedure used was to calculate the average annual cost per kW of electricity for the new nuclear power station, and to subtract from this the savings per kW which would result from *not* using older, less efficient, coal-fired power stations to provide the same electrical output; that is

> **net effective cost** = (annual cost per kW output of running the nuclear power station) −
> (savings per kW of output from not running a coal-fired power station)   (7.3)

If the savings are greater than the cost of providing and running the new power station, then the net effective cost is negative and, according to this economic appraisal model, investment in the new nuclear plant can be justified.

The components of the cost of running a nuclear power station are:

(i)   repayment of capital costs (including interest);

(ii) fuel costs;

(iii) general running costs (that is, operation and maintenance);

(iv) the cost of decommissioning the reactor.

Answering Question 7.2 will allow you to study an example of the application of this method of economic appraisal.

Question 7.2 The average annual components of the net effective cost for a 1 100 MW nuclear power station over the operational life of the reactor are as follows (in £ million yr^{-1}):

repayment of capital	99
fuel	38
operation and maintenance	17
decommissioning	12
saving from not running the fossil-fuelled plant	181

What is the net effective cost of the nuclear power station (in pounds per kilowatt per year), and can investment in it be justified?

This way of making an investment decision depends on a number of key factors. In the case of a PWR these include the lifetime and capital cost of the station (because this affects the size of the average annual repayment of the capital costs).

In addition to being sensitive to all the factors affecting discounted cash flow analysis, the value of the net effective cost will also depend on the fossil fuel prices that are assumed over the lifetime of the nuclear power station. We have already seen that long-term predictions are generally fraught with difficulties; predicting fossil fuel costs is no exception, as we shall see later. There is thus ample scope for error and disagreement.

Summary of Section 7.3

1 *Discounted cash flow analysis* is a technique that an organization can use when trying to predict and compare the profitability of future long-term projects. As used here, it works with figures that have been corrected to an inflation rate of zero.

2 The positive and negative cash flows over the lifetime of the project are converted to their positive and negative present values, the values they would have at the start of the project. To make these calculations a discount rate has to be estimated; this is an interest rate that might be earned by alternative investment.

3 If the sum of the present values, the *net present value*, is positive, the project is deemed to be more profitable than the alternative investments.

4   Another way of using discounted cash flow analysis to compare project profitabilities is to compare internal rates of return. The *internal rate of return* is the discount rate that yields a net present value of zero.

5   In justifying the PWR reactor at Sizewell, the CEGB used the method of *net effective cost*. This subtracts the savings that arise from not using older, less efficient plant, from the cost of new plant.

7.4 Research and development costs of nuclear power

Modern industries incur large research and development costs. If the government pays some of these costs for a particular industry, then that industry is receiving a subsidy, and its accounts will not show the *hidden costs* of research and development work. This is the situation with nuclear power. In the UK, much research and development work for the nuclear industry has been carried out by the government-owned UKAEA (now called AEA Technology), which has research centres at Harwell and Culham (Oxfordshire), Winfrith Heath (Dorset), Risley (Lancashire) and Dounreay (Caithness). The UKAEA's expenditure in the 30 years from 1958 to 1987 is shown in Figure 7.4.

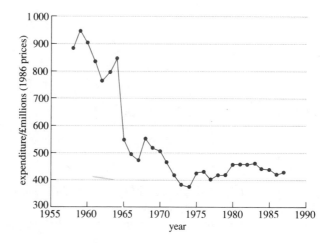

Figure 7.4   The annual expenditure of the UKAEA between 1958 and 1987 on both nuclear fission and nuclear fusion programmes, given in real terms for 1986.

▷   From Figure 7.4, how has UKAEA expenditure varied with time? What reasons can you attribute to this variation? (If you are not sure what the phrase 'in real terms for 1986' means you should read Box 7.2.)

▶   The expenditure halved in 1986 money terms between 1958 and 1972; thereafter it remained about the same level up to 1987. This is because, as the nuclear power programme became operational, there was less need for research and development.

Box 7.2   Expressing money in 'real terms'

7.2

Because of inflation, it can be very misleading to compare *actual* expenditures in different years. As you saw in Section 7.3, inflation is one reason why the value of a particular sum of money decreases with time: as time passes, the sum of money buys progressively smaller amounts of labour or goods. For this reason, comparisons are normally made in terms of the value of money in a *particular year*; amounts in other years are then increased or decreased by a factor proportional to the inflation over the period concerned.

Money that has been 'corrected' in this way is said to have been evaluated in **real terms** for the year concerned. ■

Because research and development was funded by the UK government (mainly through the former Department of Energy), the expenditure on it does not feature in the nuclear power costings given by the former CEGB. Let us see what difference it would make if it had been.

Question 7.3 In 1989 the cost of electricity to domestic consumers was $6\,\mathrm{p\,(kW\,h)^{-1}}$. In the first annual report for Nuclear Electric plc, it states that in 1989–90 the nuclear power stations in England and Wales produced 42.5 TW h of electricity. Assuming that the level of expenditure by AEA Technology in 1989–90 to be similar to the average since 1975 (Figure 7.4), and also that in 1989–90 one-quarter of this expenditure was on work relating to the current nuclear power programme, what were the 'hidden costs' of electricity from nuclear power, expressed in terms of (a) pence per kilowatt hour, (b) as a percentage of the price paid by the consumer? (*Note* $1\,\mathrm{TW} = 10^{12}\,\mathrm{W}$.)

You may wish to question the assumptions we made here, or to try to allow for expenditure made in earlier years; the possibilities are endless. However, the key issue is a question of principle, which is by no means unique to nuclear power: *if a government decides to invest in an activity for what it considers to be the good of the country, or a group of people within it, should they expect to recover the money and, if so, how?*

As with other questions that we pose, we offer no answers. It is for *you* to decide.

Figure 7.4 shows that in 1987, the UKAEA's research and development expenditure was about £440 million. Of this, fast-breeder reactor development took 17% and was the largest single item. As we have already emphasized, fast-breeder reactor development is important because it has the potential to make uranium an energy source comparable in scale with coal.

In Figure 7.5, the government investment in fast-breeder reactors is compared with that of another long-term nuclear option, nuclear fusion (Chapter 9), and with expenditure on renewable energy sources and energy efficiency. Clearly, expenditure on fast-breeder reactor development has been much greater than on all of the other three areas, although this difference is decreasing. (In fact at the time of writing (1993) the British government was proposing to withdraw from all international collaborative research on fast-breeder reactors.)

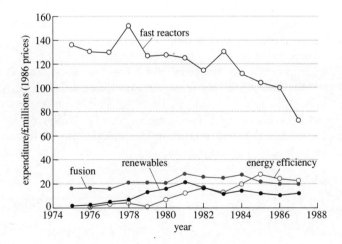

Figure 7.5 Comparison between government funds allocated to different energy-related projects between 1975 and 1987 (expressed in real terms for 1986).

There are two possible interpretations of this investment pattern:

(a) In the 1960s and 1970s the government put money into the technology that appeared to offer the greatest long-term potential. Many of the problems relating to fast-breeder reactors have now been solved, and it is possible to divert resources to other areas of energy research.

(b) The very heavy investment in fast-breeder reactors was the result of a powerful pro-nuclear lobby, which distorted the energy investment pattern.

Again, we do not propose to try to decide between these two views. The problems of competition for limited funds are almost universal, and the question of how to assign priorities is constantly under revision. Instead we simply pose this question: *how should a government decide on its allocation of resources in areas where there are several widely different activities, each at a different stage of development and with different potentials?*

This question should exercise you long after you have finished this Course.

Summary of Section 7.4

1   In the UK, a substantial proportion of the research and development costs of nuclear power were borne by the government through the UKAEA. These 'hidden costs' of nuclear power were therefore covered by a government subsidy.

2   Between 1975 and 1987, a high proportion of government investment in energy research was devoted to fast-breeder reactors rather than to energy efficiency, renewable energy sources or nuclear fusion, although that proportion has recently declined substantially.

7.5 The comparative costs of nuclear and fossil-fuelled power stations

Comparisons of the cost of generating electricity by nuclear power and by fossil fuels —coal, gas and oil—show major differences in the relative proportions of the different components of the total cost. (The components were listed in Section 7.3.3.) The first difference is in the *capital costs* involved; the cost of building a nuclear power plant (about £2 000 million for a PWR at 1988 prices) is about twice that for a coal-fired station of a similar output. The second difference is that the *construction times* for coal-fired stations are shorter, mainly because they are simpler, and there is considerable experience in building them. This means that the fraction of the capital cost arising from interest that has to be paid during construction of fossil-fuelled plants is less, and any construction delays have a smaller effect on the overall cost of electricity. Overall, repayment of capital costs accounts for about two-thirds of the electricity production cost per kW h for nuclear power stations, compared to roughly a quarter for fossil-fuelled ones. Because the capital component of the cost of electricity from a nuclear power station is such a large one, it pays to run the station as much as possible. The cost of building the station has to be met, whether it runs or not: not running it therefore costs almost as much as running it. For this reason, nuclear power stations are used to provide part of the base load (Section 7.1.1).

Another major difference is in the *fuel costs*. In a fossil-fuelled station, fuel costs typically comprise 65–70% of the total, in contrast to about 15% for a PWR.

Finally, the *costs of decommissioning* are very substantially higher for a nuclear power plant than for a fossil-fuelled one, although they are incurred much longer after the end of the plant's useful life. The decommissioning of a nuclear reactor is a lengthy process. First the final load of spent fuel will be removed from the core and disposed of as discussed in Chapter 5. Auxiliary buildings and plant can be removed fairly soon after shutdown, but the reactor core and shield will be highly radioactive. These will therefore be left for a decade or more to allow the shorter-lived radioisotopes to decay. The reactor will then be taken apart and the radioactive materials disposed of. The current intention is that all sites will be returned to a 'green field' condition.

Because of these differences in the component costs of nuclear and fossil-fuelled power stations, the *comparative* total costs can be very sensitive to the different assumptions that have to be made. The high capital cost of nuclear power is very sensitive to interest rates, to the period over which loans have to be repaid and to the time it takes to build the power station. In contrast, although factors like interest rates and repayment times also clearly affect the cost of electricity from fossil-fuelled power stations, the impact on the total cost is not so great. This is because, as noted above, fuel costs rather than capital costs are the major expenditure in this case. It is therefore possible to bias predicted costs in favour of one type of power station or another, according to the assumptions that are made regarding these parameters.

▷ Which type of power station will have costs that are more sensitive to delays in construction?

▶ The cost of electricity from nuclear power plants will be more sensitive to construction delays than that from fossil-fuelled plants, because there will be a greater increase in the *interest during construction*. Hence, accurate estimation of building times is more important in nuclear power costings. Such estimates are also more difficult, since the technology is less standardized, especially in the UK.

▷ What will be the main problem in estimating the costs of electricity during the lifetime of fossil-fuelled power stations?

▶ Because the main component of the cost is the fuel, it will be the prediction of how fossil-fuel prices will vary *in real terms* over a 20–30 year period, which is an extraordinarily difficult task.

Finally, let us consider decommissioning and waste disposal. In Activity 7.2 you considered the case of a reactor that is decommissioned after 50 years at a cost of £1 000 million. This is the sort of figure that has featured in the PWR inquiries we have mentioned. You saw how, because of the long delay, the present value of this cost and its impact on the cost of electricity is very sensitive to the discount rate used.

Apart from the uncertainty over what discount rate to use, and over the time after which decommissioning is to be undertaken, there is also much uncertainty over the actual decommissioning costs. This is because of a lack of practical experience: so far, no large nuclear power reactors have been decommissioned, so there is ample scope for disagreement. What *is* generally agreed, however, is that both Magnox and AGR reactors will cost more to decommission than PWRs, because of both their size and the fact that they have graphite moderators. There will be a much greater amount of work in dismantling them, and more radioactive waste storage space will be required.

Summary of Section 7.5

1 The chief differences between the costs of nuclear and fossil-fuelled power stations are the costs of (i) construction, (ii) fuel, and (iii) decommissioning.

2 For fossil-fuelled stations, the largest cost component is due to the fuel, whereas for nuclear power stations the capital cost component is the dominant one. As a consequence of these differences, comparisons of estimated costs for the two types of power station are sensitive to the economic assumptions made.

3   The costs of decommissioning and waste storage are influenced by (i) the actual cost, (ii) when it takes place, and (iii) what discount rate is assumed if the cost is to be met by investment. The absence of any experience of the decommissioning of large reactors, combined with the long delay before it will be complete, make these costs very hard to estimate.

7.6 Different views on the costs of nuclear power

As Section 7.5 showed, it is extremely difficult to estimate the future costs of building, running and decommissioning a nuclear power station. Very different costings may therefore be put forward, and it may not be easy to obtain agreement. Thus, the Fourth Report of the House of Commons Energy Committee (1990) says:

> *On the basis of the evidence which we have received, we are convinced that there has been a systematic bias in CEGB costings in favour of nuclear power... We believe that the Department of Energy, as the CEGB's sponsoring department, must share the blame for this, since it apparently made no attempt to obtain realistic costings from the CEGB until it was seeking to privatize nuclear power.* (para. 48)

but, later in the same report they say:

> *We have no evidence that they [the CEGB] deliberately misled the Department [of Energy] … however, they appear to have misled themselves until the onset of privatization injected more rigour into their costings.* (para. 110)

However, in a long and very critical analysis Professor Colin Robinson gives some insight into how this state of affairs might have arisen. In *The Power of the State: Economic Questions over Nuclear Generation* (Adam Smith Institute, 1991) he says:

> *The public was misled. But that was not so much a matter of evil intent on the part of individuals as of the regime under which the electricity supply and nuclear industries operated … In the case of British nuclear power, the persistent over-optimism about nuclear costs appears to have stemmed from various forms of monopoly and from the politicization and centralization of decision making … That is not to say that they deliberately distorted the estimates which they produced. 'Appraisal optimism' by those who champion particular causes is a common enough phenomenon in most industries.* (p. 17)

And he observes in a similar vein:

> *The CEGB operated in a highly politicized regime in which it was not permitted a free choice of generation fuels and in which it could pass on to consumers virtually any costs it incurred. Successive governments used it as an instrument to support the British coal industry and the British nuclear industry.* (p. 18)

Other commentators have made harsher judgements. Thus, in *The Economic Failure of Nuclear Power in Britain* (Greenpeace, 1990) Alex Henney writes:

> *The [UK]AEA and CEGB must hold the record for more wrong decisions over a longer period of time and at greater cost to the nation than any other civil organization in the country. None of the [UK]AEA's reactors paid off; the CEGB's standard of commercial decision making has been disgraceful, and at times it behaved with at best a limited sense of financial responsibility. Their officials have produced a welter of opinionated, fickle, facile and not infrequently manipulated figures and views about nuclear power.* (Summary, p. 4)

Not surprisingly, such accusations are strongly denied. For example, Lord Marshall, the former Chairman of the CEGB, writing in *National Power News* (December 1989—in part of the same article as Extract 7.2) said:

> The CEGB has never knowingly misled the public or public inquiries about nuclear costs or nuclear electricity prices. It is also total nonsense that the CEGB has not kept Ministers and the Department of Energy properly informed. In 1983 and 1985 the CEGB made clear—in publications entitled Analysis of Generation Costs—that changes in fossil fuel prices had made Magnox marginally more expensive than fossil-fuelled stations. In 1988 I told the British Nuclear Energy Forum (in a lecture which was printed in full) that Magnox stations had become significantly more expensive primarily because of reprocessing and waste treatment costs. I also said that Hinkley Point B [an AGR], which we had reported as a bit better than Drax first half [the first unit of a large, coal-fired station] was now a bit worse and that the old AGRs—Heysham, Hartlepool and Dungeness—were still having difficulties. (p. 4)

Having now looked in a general way at the problems associated with estimating the cost of nuclear power, we shall now consider how these uncertainties are reflected in the cost of electricity from a particular type of nuclear power station.

7.6.1 Comparing the cost of electricity from PWRs with coal-fired power stations

Because they represent the most likely future nuclear power source, we shall concentrate on the costs of PWRs, which were extensively discussed at the 1988 inquiry into plans to build a PWR at Hinkley Point, Somerset.

At the Hinkley Point inquiry the CEGB gave a figure of $2.24\,\mathrm{p\,(kW\,h)^{-1}}$ at a 5% discount rate for the cost of electricity from a PWR, which they compared to that for a coal-fired station of $2.62\,\mathrm{p\,(kW\,h)^{-1}}$. The main criticisms made by different groups opposed to the project were that:

(i) these initial cost estimates did not include interest during construction (Section 7.2);

(ii) the discount rate assumed was too low (Section 7.3.1);

(iii) the load factor assumed (70–75%) was too high (Section 7.1.1);

(iv) the repayment period for the capital borrowed (40 years) was too long;

(v) the estimated construction costs were too low.

The effects of making changes in the costings to meet these objections are set out in the Fourth Report of the House of Commons Energy Committee (1990):

(i) including interest during construction at 8% adds $1.09\,\mathrm{p\,(kW\,h)^{-1}}$;

(ii) increasing the discount rate (which as noted in Section 7.3.2 is set by the government for nationalized industries) from 5 to 8% adds $0.85\,\mathrm{p\,(kW\,h)^{-1}}$;

(iii) reducing the load factor from 70–75% down to 60% increases the cost by $0.7\,\mathrm{p\,(kW\,h)^{-1}}$;

(iv) decreasing the capital repayment period from 40 to 20 years adds $0.75\,\mathrm{p\,(kW\,h)^{-1}}$;

(v) increasing the construction costs by between 20 and 30% increases the cost of electricity by 0.5–$0.8\,\mathrm{p\,(kW\,h)^{-1}}$.

If all these criticisms were valid, the cost of electricity from a PWR would be increased by between 3.89 and $4.19\,\mathrm{p\,(kW\,h)^{-1}}$ from the $2.24\,\mathrm{p\,(kW\,h)^{-1}}$ claimed by the CEGB. Of course, the price of electricity from fossil-fuelled power stations would also increase if similar changes were made in the assumptions about capital costs, but to a lesser extent because, for fossil-fuelled stations, the main cost component is in the fuel. Thus, changes in the assumptions change the conclusions. The charge of bias arises because it is claimed that the CEGB consistently used assumptions that favour nuclear power.

The same Energy Committee report also criticized the CEGB's assumptions about the cost of fossil fuels:

> *Another major factor affecting the economics of nuclear power is that the costs of fossil-fired generation have fallen appreciably, mainly due to lower world market prices for fossil fuels in the wake of the collapse in the oil price from early 1986.* (para. 15)

In criticizing the CEGB's estimates Alex Henney said (*The Economic Failure of Nuclear Power in Britain*, Greenpeace, 1990):

> *The Board [the CEGB] assumed that North Sea gas would become scarce and increase in price, and that low productivity increases and continuing real wage increases in British Coal would result in a continual increase in the average pithead price of coal from 155 p GJ^{-1} to 215 p GJ^{-1} in 2000, which was then inflated to 260 p GJ^{-1} by adding transport costs. The trend in pithead prices misrepresented the post war historical pattern of coal prices, which were constant at about 100 p GJ^{-1} (March 1982 prices) from 1948–49 until 1974–75, when they increased rapidly to 170 p GJ^{-1} in 1980–81, and subsequently have declined to 123 p GJ^{-1} in 1987–88.* (p. 121)

At the public inquiry into the PWR at Sizewell, there were similar discussions over coal prices. A summary of these appears in *Critical Decision: Should Britain Buy a Pressurized Water Reactor?* (Friends of the Earth Trust, 1986), which also gives the effect of the differences in assumed coal price on the CEGB's estimates of the net effective cost (Section 7.3.3) of the Sizewell B reactor.

The above Friends of the Earth document summarizes the figures given by the various participants in the inquiry for their estimates of the increase in coal prices, and the effect that these estimates would have on the CEGB's net effective cost. Thus, in their submission, the CEGB postulated a 30% increase in the real price of coal between 1983 and 1990, a 90% increase by 2000 and a 180% increase by 2030. They estimated the net effective cost of the Sizewell B project to be −£83 kW^{-1} yr^{-1}; that is, there would be a net saving in constructing it.

On the other hand, the Stop Sizewell B Campaign suggested that the increase in real terms of the price of coal would be much less than this. They estimated that coal prices would be constant until 1990, and then increase by 1% annually until 2030, giving a 10% increase by 2000 and a 40% increase by 2030. By using their estimates, the value of the net effective cost is reduced by £66 kW^{-1} yr^{-1} to £−17 kW^{-1} yr^{-1}, which is a marked reduction in the savings attributable to the building of Sizewell B compared to the CEGB value.

So, lowering the assumed coal prices makes the net effective cost of the PWR project less negative (that is, there are fewer savings), but there would still be a net saving if coal prices were the only factor. However, if all the other criticisms made of the CEGB's assumptions are incorporated, the net effective cost of the Sizewell B PWR becomes positive, although how much so depends on which set of figures one takes.

The problem of forecasting fossil fuel prices, which are affected by unpredictable domestic and international events, is not a new one. In *The Power of the State: Econ-*

omic Questions over Nuclear Generation (Adam Smith Institute, 1991), Professor Colin Robinson says:

> *Almost everyone has also been very bad at forecasting the prices of fossil fuels. Unless, therefore, it is assumed that forecasting ability has miraculously improved, any cost comparisons between nuclear and fossil-fuelled generating plant have to be viewed with the utmost caution. They are little more than informed guesses...* (p. 87)

Summary of Section 7.6

1 It has been claimed that, prior to privatization, the difficulties of estimating the future costs of building, running and decommissioning power stations allowed the CEGB to make assumptions in their economic appraisals which favoured nuclear power.

2 There were mitigating circumstances: the CEGB was not operating in a free market, and was under government pressure to support both nuclear power and the British coal industry.

3 In the case of the PWR, critics claim that the CEGB's economic appraisal failed to include interest during construction, and assumed unduly low values of the discount rate and construction costs, and unduly high values of the load factor, the capital repayment period and the increase in the real price of coal.

7.7 A concluding activity

We have given a number of examples of the CEGB's estimates of nuclear power costs, and of the criticisms that have been made of them. In doing so, our intention has been to make you aware of the issues, rather than provide you with enough detail to resolve them to your own satisfaction. What we have not done, yet, is to give any details of the CEGB's counter criticisms and comments. As an example of one of them, we shall therefore conclude this chapter with an activity based on part of the text of a lecture given by Lord Marshall (the chairman of the CEGB before it was privatized) to the British Nuclear Energy Society in November 1990, in which he discusses the issues surrounding the proposed privatization of nuclear power (which focused attention sharply on the cost of nuclear power).

Activity 7.3   *You should spend up to 30 minutes on this activity.*

Extract 7.2 is the section of Lord Marshall's speech relating to PWRs. Read it and then explain in a sentence or two:

(a) the problems that gave rise to the series of government decisions relating to nuclear power;

(b) the main factors that led to an increase in the estimated cost of electricity from PWRs;

(c) by what percentage the estimated price of electricity rose as a result of having only a 20-year contract to supply electricity.

Extract 7.2   From *National Power News*, December 1989.

a) structure of industry } useful
Policy (for competition)

fuer.

The facts about nuclear power

BY LORD MARSHALL. *This article is based on his lecture, The Future for Nuclear Power, delivered to the British Nuclear Energy Society in the Royal Lancaster Hotel, London on November 30.*

In the light of the Government decision to abandon the privatisation of nuclear power I would like to set out for the record an explanation of what has happened in the last two years.

The Government has made three basic decisions. In July they decided to maintain the Magnox stations in the Government sector.

Next they decided to abandon the full PWR construction programme and finally they decided there was no purpose in privatising the AGRs.

These decisions are the consequence of two major problems—the first institutional, the second financial—that have emerged from the Government's privatisation proposals. In the event these problems made it impossible for nuclear power to survive in the private sector.

The institutional problem concerns the structure of the industry and the overriding priority given to competition, leading to the abandonment of the obligation to supply.

The financial problem concerned the rate of return on investment required in the private sector and the position of the investing institutions and banks on risks and uncertainties. This would have put up the costs of nuclear power in the private sector to unacceptable levels in the absence of considerable underwriting by Government.

The financial implications affected the different reactor types in different ways and I would like first to discuss the institutional problem.

In my discussions with Cecil Parkinson late in 1987 and early in 1988, I argued that a successful nuclear power programme is best pursued by a large generator with the obligation to supply in a defined geographical area.

For all practical purposes the CEGB has the franchise for England and Wales with the ability to recover its costs from customers (through the area boards) in return for the obligation to provide reliable supplies in real time and in the indefinite future. That future obligation was the driving force for the CEGB to seek to diversify its fuel and plant base and to develop nuclear power.

The Government White Paper dramatically changed these arrangements. The CEGB was to be split into two generators and a separate grid company. Neither generator would have a franchise nor the ability to pass costs through to the ultimate customer.

The obligation to supply would be transferred to the 12 area boards, to be called distribution companies.

But the Government also wished to introduce competition into distribution which gave sharp conceptual difficulties.

Should the distribution company plan to supply all customers in its area or only those customers it thinks it will get? If it thinks it will get only some of the customers in its own area, what is its obligation to plan the long-term future for supplies in that area?

The electricity industry has sought an answer which would preserve both competition and the security of supply but we have finally come to the conclusion that no workable compromise is possible.

This issue, which has hung over privatisation for 18 months, was finally put beyond further debate by John Wakeham in September when he confirmed that competition was to be the guiding principle.

Since that was incompatible with the obligation to supply, then after an interim period the distribution companies would have no obligation to supply in the traditional sense. They would, of course, have an obligation to connect new customers, and they would have an obligation to make a reasonable offer to supply electricity as soon as they could acquire it.

But the precise amount of generation available at any particular hour, on any particular day, in any particular year in any particular decade would now be determined by market forces.

The distribution companies will now face some competition immediately, an increased degree in four years' time and total competition in eight. They cannot know what the extent of their business will be. How then can they willingly give lifetime contracts to buy nuclear power?

If a private National Power is to build a nuclear station, a long-term contract, preferably a lifetime contract, to sell the electricity is needed. Without that contract National Power cannot raise the money from the banks.

It follows immediately that National Power cannot build a PWR programme.

The plain fact of the matter is that we are going to have an electricity industry driven by short-term market considerations and fierce competition. You cannot introduce nuclear power because the benefits, assuming we get the technology right, accumulate over half a century. Similar arguments, with a shaded emphasis, apply to large coal-fired plants.

It is my belief that the Government was faced with a stark choice between the long-term benefits of nuclear power and the short-term benefits of privatising the industry in this particular form. They chose privatisation.

The PWRs offer many answers

The crucial question about the PWRs is the price of electricity from these reactors in the private sector and in the public sector. Commentators assume there is only one answer. There is in fact an infinite series of answers to both.

The price is the sum of four parts: the capital charges (which include profits), operating costs, overheads and deferred charges such as decommissioning.

Overheads include the cost of head office, rates and insurance, and a variety of matters of that kind which, roughly speaking, are constant for all power station types.

Traditionally they have not been included in figures given at public inquiries or in analyses of comparative generating costs because they are not a "new resource cost" and not necessary for comparing the value of one power station against another. They are, anyway, not dominant and for a PWR amount to 10 per cent of the total price.

The operating costs of a PWR, including the fuel, are of course important and we estimate they amount to about 15 per cent of the total price.

The factor which overwhelmingly determines the price of electricity is the charge on the capital. This is 75 per cent of the price of electricity and consists of two terms—the return earned on the capital and the depreciation on the plant.

As a public sector company the CEGB set the return on capital on the recommendations of the Treasury. Since 1977 they have advised us that return on capital should be five per cent. Last year they raised that to eight per cent.

If the CEGB depreciated the plant over its full design lifetime, 40 years, we would get an average price of electricity over 40 years of 3.2 pence a kilowatt hour.

In the private sector finance the answers are different. So how do we construct a realistic case?

We need to start with the contract with our customers. Suppose we could get a 20-year contract with them. Then we must depreciate the plant over 20 years because, although the plant might be technically all right for the long term, we cannot be certain that we can earn money from it.

The distribution companies would argue that we should accept the risk of construction and the risks of non-performance. In practice, therefore, we need to see what risks we and our shareholders must accept and in consequence what rewards we should seek to compensate the shareholders for the risk element in these contracts. We had best be more prudent at guessing the availability of the plant—else we are certain to make a loss.

Above all else, if we can only borrow money for 10 years, we need to make very high charges early to enable us to repay the debt. On the other hand, if we borrow money over a long period of time then we need strong guarantees either from the distribution companies or from the Government.

Somewhere amongst all these choices we must find a package that is acceptable to everyone. There is, in fact, a broad range of answers—though none of them as extreme as some reports in the press suggest—which relate principally to the risks to which a privatised National Power might be exposed in a number of areas and to the conditions which banks or investors might seek before committing funds.

We accepted that National Power was not being offered a totally risk-free environment in which to sell its product. We were not going to be able to pass on all its costs whatever they might be and regardless of plant performance and earn a predictable profit. The bankers may or may not have gone along with this in the event.

On October 11 we finally gave Government a range of indicative prices for the period up to the end of the century plus a possible settled down price. That was 6.25 p/kW h.

The factors which push up our estimates of what prices might have to be were the realities of satisfying the financial markets.

The prospective operating life of a PWR is about 40 years and we had major doubts as to whether we could negotiate contracts with distribution companies which would hold for 40 years and give us confidence that we could recover our fixed cost burden. It seemed improbable that Government devices such as the non-fossil obligation and the fossil fuel levy could be made to hold that long.

But, more significantly, we were getting messages from the banking community with two emphases.

THE FIRST was that we would be expected to pay off our borrowings over a much shorter period than 40 years and that we should therefore seek shorter contracts. So we based our price indications on contracts which would recover all our fixed costs and capital over 20 years and which earned a real terms return of 10 per cent before tax and interest, a figure the Department of Energy had invited us to use.

Even then we were going to be in some difficulties because our interest bill was going to absorb a high proportion of the cash flow of the project in the early years and the project would show a loss.

THE SECOND MESSAGE from the banks turned out to be crucial though hardly unexpected. Banks were seeking full Government guarantees of the debt as well as assurances that all significant risks would be either passed through to consumers or carried by Government itself.

On November 9 John Wakeham gave his answer in the House of Commons when he announced his decision on the nuclear stations:

"Discussions have taken place about financing new nuclear power stations... In the event unprecedented guarantees were being sought. I am not willing to underwrite the private sector in this way."

Private sector price for PWRs	
	pence/ kWh
Public sector price (risk free)	3.22
Adjustment from 8% to 10% internal rate of return	0.71
Uncertainties: 70% availability, instead of 75% and fuel reprocessing and decommissioning uncertainties	0.54
Different basis for calculating profit (10% IRR* to 10% current cost accounting)	1.03
	5.50
Increase for 20-year contract	0.75
	6.25

What privatisation meant for PWR prices: In October 1987 the CEGB told the Hinkley C inquiry that the resource cost for the station would be 3.09 pence a unit at an 8% internal rate of return. After allowing for inflation and other changes, including allocation of overheads, that would give a current *public sector price* on a risk free basis of 3.22 pence a unit. The *private sector price* quoted to the Department of Energy is determined by adding in four main elements: an increase in the real rate of return; additional costs relating to uncertainties which the private sector would treat very cautiously; a different way of calculating profit; and an increase to recover the capital cost over 20 years instead of 40. The 6.25p/kWh is the settled down private sector price for the four PWRs.

Provisions for Magnox power stations

	£million
Total at 31 March, 1988	2 800
Fuel cycle effects (BNFL):	
Fixed price, backlog of Magnox waste, decommissioning Sellafield	3 100
Station decommissioning	900

ALL SUBJECT TO REVIEW

Increases in the back-end costs of the Magnox power stations meant that CEGB provisions to cover these costs would have to be increased.

* IRR is the internal rate of return; see Section 7.3.2

8 Nuclear weapons proliferation

In most people's minds the start of the 'nuclear age' was not the first nuclear reactor, built in Chicago in 1942, but the exploding of atomic bombs over Hiroshima and Nagasaki in 1945. The use of nuclear reactors to produce energy for electricity generation did not follow until 1954, although reactors were built earlier than this to make one of the weapon fuels (^{239}Pu).

For military reasons the governments of the countries who first developed nuclear weapons (the USA, followed by the former USSR, Britain and France) did not want possession of these weapons to spread; they would have lost their perceived military advantage. In addition, there was world-wide concern that the spread of nuclear weapons could have horrifying consequences for the whole world. For these reasons, there has always been an awareness of the need to try to separate the use of nuclear energy for electricity generation from its use for weapons; 'atoms for peace' was a 1950s slogan.

In this chapter, we shall first look at what nuclear weapons are, and what is needed to produce them. We shall then examine the relationship between nuclear power and nuclear weapons, and whether this relationship needs to exist. Finally, we shall consider the steps that have been taken to prevent the proliferation of nuclear weapons, and examine the case of Iraq to see how effective these measures have been in one particular case.

8.1 The principles of nuclear weaponry

An explosion is a very rapid release of energy—that is, occurring in a fraction of a second. In 'conventional' explosives this is achieved through *chemical* reactions, and, for a particular reaction, the amount of energy released is determined by the mass of the explosive involved. Trinitrotoluene (TNT) is a widely used, and powerful, explosive and, as you probably know, explosions are often compared in terms of the mass of TNT needed to produce the same amount of energy.

Nuclear weapons, on the other hand, work on very different principles, even though the energy released is still often quoted in terms of the equivalent mass of TNT. The fundamental difference is that the source of the energy is a *nuclear* reaction, either fission or fusion.

▷ What is a fissile isotope?

▶ A fissile isotope is one that undergoes fission with neutrons of any energy (Section 2.3.1). ^{235}U is a naturally occurring example, and ^{239}Pu is one that can be produced in nuclear reactors.

▷ What is the characteristic of a supercritical system?

▶ It is one in which the number of fissions occurring per unit time (the fission rate) is increasing (Section 2.5.3).

A **fission weapon** is a supercritical assembly of pure, or nearly pure, fissile material in which very large amounts of energy arising from fission are released in a few microseconds.

In a **fusion weapon**, on the other hand, a fission bomb is used to provide the energy to heat up and compress a mixture of deuterium and tritium ($^2_1$H and $^3_1$H, respectively). The main energy release comes when these two isotopes fuse together; we shall discuss the peaceful application of this reaction in Chapter 9. Because deuterium and tritium are both isotopes of hydrogen, such weapons are sometimes called 'hydrogen bombs'.

Figure 8.1 shows a *schematic* cross-section of a fission bomb. The main component is a spherical shell of nearly pure fissile material, either ^{239}Pu or uranium enriched to over 90% in ^{235}U. The mass of material is sufficient to make a supercritical assembly, but the shell into which it has been fashioned is initially *subcritical* because it has been made in a porous form with a low density. A *supercritical* assembly is produced from this shell when it is compressed by conventional, chemical explosives which are placed around the outside of the shell. Around the fissile material is a solid shell composed of a substance that has a low tendency to absorb neutrons, such as beryllium. This then acts as a *neutron reflector*; that is, it reflects back neutrons that would otherwise escape and be lost from the chain reactions.

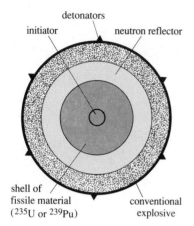

detectors

initiator   neutron reflector

shell of fissile material (^{235}U or ^{239}Pu)   conventional explosive

Figure 8.1   Schematic section through a nuclear fission weapon.

One of the important features of nuclear weapons is that as much energy as possible should be produced before the weapon itself is blown apart. For this reason, it is important to establish as many chain reactions as possible, as quickly as possible. Since it is neutrons that initiate chain reactions, this is achieved by putting a source of neutrons, the **initiator**, at the centre of the weapon. (When they are started up, nuclear power reactors also use neutron sources to initiate the chain reaction.) Neutrons can be produced by the nuclear interaction between α-particles (which you will recall are helium nuclei) and beryllium:

$$^9_4\text{Be} + {}^4_2\text{He} \longrightarrow {}^{12}_6\text{C} + {}^1_0\text{n} \tag{8.1}$$

One source of α-particles is the radioisotope polonium-210.

▷   In what context have we mentioned this isotope before?

▶   This isotope was being produced in the Windscale reactor when it caught fire (Section 6.2.1). Release of polonium-210 accounted for some of the deaths attributed to the accident.

In order that the neutrons from the initiator should be produced only when they are needed, the polonium-210 is covered by a thin layer of gold, which prevents the α-particles from interacting with the beryllium until the conventional explosives are fired. The detonators for the conventional explosive are evenly spaced over the surface of the weapon, and all have to fire at the same time (within millionths of a second or less).

In Chapter 6, we said that the *chemical* energy released in the Chernobyl explosion was equivalent to that from 100–$200\,kg$ of TNT. The only two nuclear weapons ever used in a war (those dropped on Hiroshima and Nagasaki) were both fission devices, one using ^{235}U and the other ^{239}Pu, and the energy that they released was equivalent to $12\,500$ and $20\,000$ *tonnes* of TNT, respectively.

Although the weapons used on Hiroshima and Nagasaki were small compared to those that have subsequently been produced (a fusion bomb equivalent to 50 *million* tonnes of TNT was tested by the former USSR), their destructive power was, nevertheless, horrifying, even by the standards of modern warfare. As noted in Section 4.5.2, more than 50% of the inhabitants of both cities died either immediately or within 5 years.

8.2 Production of fissile isotopes

If a country wishes to build nuclear weapons, the first requirement is to obtain nearly pure ^{235}U or ^{239}Pu; in both cases 'nearly pure' means 90% or more of the isotope concerned.

▷ How can these isotopes be produced?

▶ Nearly pure ^{235}U has to be produced by enrichment (Section 5.2.1); ^{239}Pu is produced in a nuclear reactor following neutron absorption by ^{238}U (Equation 2.12).

So, a country aspiring to be a nuclear weapons power needs to have either a uranium enrichment plant or a nuclear reactor.

▷ What else is needed to make a weapon from plutonium?

▶ A fuel reprocessing plant to extract the plutonium from the unused uranium and the fission products (Section 5.5).

The amount of plutonium needed to make a nuclear weapon depends in part on the skill of the weapon designer, and in part on the isotopic composition of the plutonium. The higher the content of the fissile isotope, ^{239}Pu, the smaller (in size) the weapon can be. The fact that you will see a range of masses quoted—from a few kilograms up to 20 kg or so—reflects the different isotope content and weapon designs. A mass of 10 kg would be a reasonable value to use for discussion purposes. Because uranium and plutonium are about three times as dense as steel, 10 kg of fissile material occupies very little space. The diameter of the spherical shell of fissile material in an actual weapon would not be much greater than 15–20 cm or so.

Although ^{235}U was used in the first nuclear weapon, its production through enrichment is technologically difficult, and production rates are low (Section 5.2.1). In contrast, the production of ^{239}Pu in a nuclear reactor is relatively rapid; roughly 1 g per day would be produced in a reactor producing 1 MW of heat energy. Thus, a reactor producing, say, 25 MW of heat energy would produce the 10 kg of ^{239}Pu which is needed for a nuclear weapon in about 400 days. This leads to an important conclusion: the large nuclear power reactors that are used for electricity production and produce some 3 000 MW of heat energy (Section 3.2) *are not necessary in order to produce plutonium for use in weapons*. Quite small reactors, not designed to produce electricity, can be used. In fact, you will remember that the Windscale reactor that caught fire (Section 6.2) produced plutonium without making electricity; the reactor was air cooled, and the hot air was simply pumped out into the atmosphere.

8.2.1 The production of ^{239}Pu

There are currently six countries which have declared that they have nuclear weapons (China, France, India, UK, USA and the former USSR), all of which also have civil nuclear power programmes. Israel is believed to have them, but does *not* have a nuclear power programme. The basis of the Israeli nuclear weapons programme, which is discussed in some detail by Seymour Hersh (*The Sampson Option*, Faber & Faber, 1991), is a 24 MW reactor at Dimona, in the Negev desert.

We have said that ^{239}Pu of over 90% purity (and, ideally, over 95%) is needed to make a nuclear weapon. Let us recall what we know about the plutonium from power reactors.

▷ Is the plutonium from power reactors normally pure ^{239}Pu?

▶ No, it normally also contains ^{240}Pu, ^{241}Pu and ^{242}Pu, which are formed from ^{239}Pu by neutron absorption (Figure 5.8).

Besides diluting the concentration of the fissile isotope ^{239}Pu, these higher isotopes of plutonium can cause other difficulties. For example, significant amounts of ^{240}Pu are unwelcome because this isotope undergoes *spontaneous* fission (Section 2.2.1); that is, it does not need neutrons to initiate fission. Because the neutrons produced during the spontaneous fission of ^{240}Pu will be given off all the time, these neutrons can induce fissions before the ideal time for detonation, thereby reducing the explosive yield from the weapon.

Exactly what percentage of the higher isotopes is present depends particularly on how long the fuel has been left in the reactor.

▷ If the fuel is left in the reactor for a longer time, would you expect the percentage of the higher isotopes to be larger or smaller?

▶ It will be larger: a longer stay in the reactor will increase the chance that the ^{239}Pu initially produced will successively absorb neutrons to give ^{240}Pu, ^{241}Pu and ^{242}Pu.

So if highly pure ^{239}Pu is to be produced, the fuel must spend a relatively short time in the reactor. A more sophisticated way of putting this is to say that the **fuel burn-up** must be low. The fuel burn-up tells us the energy that the fuel has produced in megawatt-days* per tonne ($MW\,d\,t^{-1}$) by the time it is removed from the reactor. The longer the fuel stays, the more energy it can produce, and the higher the burn-up will tend to be. Typical values of the fuel burn-up in power reactors were given in Table 3.2 as $18\,500\ MW\,d\,t^{-1}$ for an AGR, and $32\,000\ MW\,d\,t^{-1}$ for a PWR. To produce nearly pure ^{239}Pu, the fuel burn-up should be about $1\,000\ MW\,d\,t^{-1}$ or less. Thus, *to produce plutonium for weapons in a power reactor, the fuel has to be taken out of the reactor well before the end of its normal lifetime.* Of course, if the object is to produce weapons, this inefficient use of fuel does not matter.

Once the plutonium has been produced, it has to be extracted from the uranium and fission products. This was discussed in Section 5.5.

▷ What complicates the chemical extraction of plutonium from the other components of the spent fuel?

▶ The spent fuel is extremely radioactive, so the reprocessing plant has to be heavily shielded, and most of the operations have to be undertaken by remote control.

Finally, of course, the fissile material has to be made into a working bomb. Although the principles we have been discussing make this step sound comparatively easy, in practice this is fortunately not the case. Nevertheless, India, which has a smaller industrial, scientific and technological base than the UK, for example, has developed its own nuclear weapons.

8.3 The 'weapons connection' with nuclear power

There is a large number of countries which have nuclear power programmes but do not have, or are not known to have, nuclear weapons programmes. They include Bulgaria, Canada, Netherlands, Germany, Japan, South Korea and Sweden. We have seen, from the example of Israel, that a nuclear power programme is not necessary for the production of nuclear weapons. To what extent, then, is there a connection between nuclear power and nuclear weapons?

---

* $1\ MW\,d \equiv 24\,000\ kW\,h$.

First of all, the obvious point is that some, although by no means all, of the expertise needed to make a nuclear weapon can be found among the people who run a civil nuclear power programme. However, the weapons connection emerges more sharply if we consider the two types of fission weapon.

If the weapon is to be made from ^{235}U, then a plant capable of enriching uranium to at least 90% ^{235}U is necessary. Enrichment plants are needed to make uranium fuel for most nuclear reactors, and if a country builds a plant for this purpose, it can also use the same technology to produce weapons-grade uranium. Nations such as Sweden, however, have no enrichment plant: they import their enriched fuel, containing only about 3% ^{235}U, from other countries, and so cannot use their nuclear power programme to make a uranium weapon. This, then, is our first conclusion: *if a civil nuclear power programme does not include an enrichment plant, it cannot be used to provide weapons-grade uranium.*

Now let us turn to a weapon made from ^{239}Pu. Here the connection with the civil nuclear power programme exists only if the country in question has a fuel-reprocessing facility. Only then can the plutonium be obtained in a pure form (Section 8.2.1). If reprocessing is available, then it is relatively easy to embark on nuclear weapons development, because, up to the point where the plutonium is made into a weapon, the same scientific and technological expertise is required for the two programmes.

So, the key to the separation of the use of plutonium in civil nuclear reactor programmes from the military use of plutonium lies in fuel reprocessing. This is our second conclusion: *if a civil nuclear power programme does not include a reprocessing plant, then it cannot be used to provide weapons-grade plutonium.*

Thus, the potential danger signal as regards nuclear weapons proliferation is the existence of either an enrichment plant or a reprocessing facility.

Summary of Sections 8.1–8.3

1 Nuclear weapons are of two kinds: in a fission weapon, a subcritical mass of a fissile isotope centred on a neutron source is made critical by explosive compression; in a fusion weapon, a fission weapon is used to heat and compress a mixture of tritium and deuterium. A fission weapon is therefore an essential feature of both types of weapon.

2 The fissile isotopes used in fission weapons consist of about 10 kg of ^{235}U or ^{239}Pu with isotopic purities of over 90%. An enrichment plant is needed to make the ^{235}U; a nuclear reactor and a reprocessing plant are needed to make the ^{239}Pu.

3 It is not necessary to have a nuclear power programme to make weapons-grade plutonium; a single small reactor is sufficient.

4 Plutonium that is over 90% ^{239}Pu is produced only at low fuel burn-up. Higher burn-up yields a higher proportion of heavier isotopes such as ^{240}Pu, which reduces the effectiveness and 'shelf-life' of the weapon as a result of spontaneous fission.

5 There can only be a connection between a civil nuclear power programme and nuclear weapons if the civil programme contains enrichment or reprocessing facilities.

Question 8.1 What misunderstanding of the methods of making nuclear weapons could the following remark be said to betray?

> *...where North Korea, last of the ideological dinosaurs, enriches plutonium.*
> (*Analysis*, BBC Radio 4, 11 June 1992)

8.4 *The control of nuclear weapons proliferation*

After the Second World War, it was thought that nuclear power would make a substantial contribution to the world's energy supplies. As a result, promotion of the peaceful uses of nuclear energy was considered to be a suitable activity for the United Nations, and a special agency, the International Atomic Energy Agency (IAEA), with its headquarters in Vienna, was established to undertake this function under the auspices of the United Nations.

International concern about the spread of nuclear weapons resulted in the establishment of a Non-proliferation Treaty on 1 July 1968; sections of this are reproduced as Extract 8.1. The responsibility for implementing the Treaty was given to the IAEA. By 1990 the Treaty had been signed by 137 countries; Iraq was one of the earlier signatories, doing so in 1969. Notable non-signatories up to 1990 were France, India, Israel, Pakistan and South Africa.

Activity 8.1

Read Extract 8.1.

(a) How is the spread of weapons and expertise from the nuclear weapons states to be prevented?

(b) Can countries withdraw from the Treaty?

(c) Is it possible for a country to undertake weapons development under cover of another activity?

(d) What, in your opinion, are the important factors that should enable the Treaty to be successful in preventing the global spread of nuclear weapons?

Extract 8.1　From the Treaty on the Non-proliferation of Nuclear Weapons.

Treaty on the non-proliferation of nuclear weapons

The States concluding this Treaty, hereinafter referred to as the "Parties to the Treaty",

Considering the devastation that would be visited upon all mankind by a nuclear war and the consequent need to make every effort to avert the danger of such a war and to take measures to safeguard the security of peoples,

Believing that the proliferation of nuclear weapons would seriously enhance the danger of nuclear war,

Have agreed as follows:

ARTICLE I

Each nuclear-weapon State Party to the Treaty undertakes not to transfer to any recipient whatsoever nuclear weapons or other nuclear explosive devices or control over such weapons or explosive devices directly, or indirectly; and not in any way to assist, encourage, or induce any non-nuclear-weapon State to manufacture or otherwise acquire nuclear weapons or other nuclear explosive devices, or control over such weapons or explosive devices.

ARTICLE II

Each non-nuclear-weapon State Party to the Treaty undertakes not to receive the transfer from any transferor whatsoever of nuclear weapons or other nuclear explosive devices or of control over such weapons or explosive devices directly, or indirectly; not to manufacture or otherwise acquire nuclear weapons or other nuclear explosive devices; and not to seek or receive any assistance in the manufacture of nuclear weapons or other nuclear explosive devices.

ARTICLE III

1　Each non-nuclear-weapon State Party to the Treaty undertakes to accept safeguards, as set forth in an agreement to be negotiated and concluded with the International Atomic Energy Agency in accordance with the Statute of the International Atomic Energy Agency and the Agency's safeguards system, for the exclusive purpose of verification of the fulfilment of its obligations assumed under this Treaty with a view to preventing diversion of nuclear energy from peaceful uses to nuclear weapons or other nuclear explosive devices.

2   Each State Party to the Treaty undertakes not to provide: (a) source or special fissionable material, or (b) equipment or material especially designed or prepared for the processing, use or production of special fissionable material, to any non-nuclear-weapon State for peaceful purposes, unless the source or special fissionable material shall be subject to the safeguards required by this Article.

ARTICLE IV

1   Nothing in this Treaty shall be interpreted as affecting the inalienable right of all the Parties to the Treaty to develop research, production and use of nuclear energy for peaceful purposes without discrimination and in conformity with Articles I and II of this Treaty.

ARTICLE X

1   Each Party shall in exercising its national sovereignty have the right to withdraw from the Treaty if it decides that extraordinary events, related to the subject matter of this Treaty, have jeopardized the supreme interests of its country. It shall give notice of such withdrawal to all other Parties to the Treaty and to the United Nations Security Council three months in advance. Such notice shall include a statement of the extraordinary events it regards as having jeopardized its supreme interests.

8.4.1 A case history: the Iraqi nuclear weapons programme

How effective a treaty is can only be judged by the results it produces. Let us study a case history, that of Iraq, which, you will remember, signed the Non-proliferation Treaty in 1969. As a signatory, Iraq's nuclear installations were supposed to be open to inspection by the IAEA officials.

Officially, Iraq has not had a nuclear weapons programme, but following the Gulf War in 1991, a United Nations resolution made it obligatory for Iraq to declare, and have destroyed, all material relating to chemical warfare or weapons of mass destruction. An IAEA team was appointed to carry out these tasks, and Iraq's attempts to build a nuclear weapon have now been revealed.

Activity 8.2   *You should spend up to 15 minutes on this activity.*

Note relating to Extract 8.2: a calutron separates uranium isotopes using the principle of the mass spectrometer.

Read Extracts 8.2 and 8.3, which give details of Iraq's nuclear weapons programme. Now write answers to the following questions.

(a) What route to a nuclear weapon does Iraq appear to have chosen? If another route was being followed, how has it now been blocked?

(b) How does the view of the effectiveness of Iraq's nuclear weapons programme expressed in Extract 8.3 differ from that in Extract 8.2? What new discovery about the technology of the programme brought about this change of mood?

(c) Does either extract imply that there is firm evidence of violations of the articles of the Non-proliferation Treaty (those printed in Extract 8.1) by Iraq, or by Western governments or companies? How effective do you think the IAEA has been in policing the Non-proliferation Treaty in this case?

Extract 8.2 From *The Independent on Sunday*, 14 July 1991.

Scientists explode the myth of Saddam's nuclear bomb

An atomic threat from Baghdad is a remote prospect, Tom Wilkie reports

With the equipment Saddam Hussein has admitted to possessing, it would have taken him nearly 30 years to accumulate enough material for one nuclear weapon, according to a physicist at Imperial College, London, who has analysed Iraq's declared nuclear capacity for *The Independent on Sunday*.

The consensus among independent observers is that, despite the international alarm, Iraq is many years away from a nuclear weapons capability. Moreover, although Baghdad appears to be trying to conceal equipment, it is inconceivable it could embark on a serious nuclear weapons production programme without it being clearly detected.

The hysteria of the past few days has arisen because neither the US government nor the US press appears willing or able to distinguish between laboratory-scale research and the huge military–industrial complex required to produce nuclear weapons.

Attention has focused on Iraq's disclosure in a letter to the UN that it had eight "calutrons"—machines which can be used to produce the type of uranium needed for a nuclear explosive. Calutron technology dates back to the Forties and was the method by which the US produced the nuclear material it needed for the bomb dropped on Hiroshima.

The starting-point for all nuclear weapons is uranium, which exists in nature in two forms: 99.3 per cent is of a type known as uranium-238; 0.7 per cent is of the uranium-235 type. Only this second form, uranium-235, will make a nuclear explosive, so one route to a bomb is to "enrich" natural uranium in the 235 isotope. Calutrons are about the least efficient method of doing so.

According to John Hassard of Imperial College, the eight calutrons which Iraq says it had before the Gulf war would have produced at most 2 g of enriched uranium a day. The minimum needed for a single nuclear explosive is 20 kg, so Iraq would have needed to run the calutrons for 10 000 days—nearly 30 years—to obtain enough for one device.

The Iraqi letter discloses that 22 more calutrons were in various stages of construction, but these would not have brought Iraq within a decade of having a bomb. Dr Hassard pointed out that the US wartime Manhattan Project built 900 calutrons at Oak Ridge in Tennessee—connected to form units known as "racetracks"—and all of these took more than 100 days to enrich enough uranium for a single nuclear bomb.

Calutrons consume huge quantities of electricity, which is eventually dissipated as heat. It is impossible to hide cooling towers or any other method of rejecting heat into the environment from satellite observations.

Oak Ridge was built to take advantage of hydro-electricity from dams built in the Thirties by the Tennessee Valley Authority. No evidence has yet been produced that Iraq has built huge new power stations that it would need to provide electricity for a uranium-enrichment plant. Further, the Oak Ridge plant was huge: it required 22 000 workers, had eight electricity sub-stations, 12 water cooling towers and more than 268 buildings of various sizes. Iraq might manage with something more modest, but it would still be obvious.

The second route to a bomb is by transmuting uranium-238 into plutonium in a nuclear reactor. Here again, there are formidable problems. The two nuclear reactors known to exist in Iraq were destroyed by American bombing.

But to focus on technology, as many have done, is to miss the point. The most important constraint on a country's obtaining nuclear weapons is manpower. It is one of the canards of our time that any physics graduate with access to a technical library could produce a nuclear bomb. It is true that the basic nuclear physics of designing the weapon is easy, but the engineering is phenomenally difficult. Far more important than theoretical knowledge is to have educated, motivated people who can think with their fingertips as much as with their brains.

Most of the leaders of Iraq's nuclear programme have disappeared in purges. Hussein Sharastani, for example, a graduate of Imperial College, was scientific adviser to the Iraqi President until 1979 when, it is believed, Israeli agents sabotaged vital components being manufactured in France for the Osirak reactor under construction in Iraq. Dr Sharastani, a Shia, was arrested in 1979 and is believed to have been executed.

Saddam himself took a direct personal interest in the nuclear programme thereafter, but a country which values ideological or religious purity over technical competence is not one likely to make swift progress towards creating mushroom clouds over the desert.

Extract 8.3   From the *Daily Telegraph*, 5 October 1991.

Iraq planned N-warhead for missiles

By Roger Highfield, Science Editor

IRAQ was planning to build a nuclear warhead, the International Atomic Energy Agency said yesterday.

Mr David Kay, chief inspector of the United Nations team, besieged in a Baghdad car park last week, said the Agency has evidence of a substantial Iraqi nuclear programme—called PC3—linked to an extensive procurement programme abroad and an effort to develop a surface-to-surface missile.

The director general of the Agency, Dr Hans Blix, denied the Agency had been a "sleepy watchdog" because it had failed to detect the £5.8 billion programme, but said efforts to strengthen safeguards would be boosted.

Iraq has so far denied developing nuclear weapons and says its research was for peaceful purposes.

The UN team, which is still analysing 25 000 pages of documents and 700 rolls of film it took during its inspection tour last week, found evidence that Iraqi plants were within 12–18 months of producing enriched uranium for nuclear weapons.

A preliminary report has already been submitted to the United Nations.

The inspectors found industrial-scale plants based on old-fashioned electro-magnetic enrichment technology.

But they believe Iraq had switched to centrifuge enrichment, probably based on gas centrifuge designs developed by Urenco, the Anglo–Dutch–German consortium.

The Agency is still trying to locate Iraq's clandestine centrifuge enrichment plant and is concerned the Iraqis may still have a cache of highly enriched uranium.

There was evidence Iraqi authorities had moved documents and equipment related to a uranium enrichment programme while Mr Kay's team was detained for four days in the Baghdad car park.

With enough enriched uranium—at least 30 lb for each warhead—the Agency estimates Iraq's nuclear weapons effort at Al Atheer—the focus of PC3—could then have produced a crude nuclear weapon.

Documents collected in Iraq contained "the names of just about every European and US electronics and engineering company," said one official. The walls of the rooms in the Iraq Atomic Energy Commission were lined with catalogues from many of them.

But companies will not be named until Dr Blix and UN chiefs discuss the matter at the Security Council next week. The investigators must determine whether firms knowingly co-operated with Iraq's nuclear programme or were unaware of how their products and expertise were being used.

Mr Rolf Ekeus, head of the UN Special Commission in charge of scrapping Iraq's weapons of mass destruction, arrived in Baghdad yesterday to warn the Iraqis not to impede the inspection teams.

The Agency estimates that the Iraq nuclear programme could be revived within five years.

The UN needs to ensure that, after all the equipment and paperwork for Iraq's nuclear weapons programme is destroyed, it cannot be restarted from information still in the heads of the 7 400 people who worked in the nuclear weapons programme.

8.4.2 Need there be a 'weapons connection'?

We have seen that Israel does not have a nuclear power programme, yet has undertaken the development of nuclear weapons by using a low-power, non-electricity producing reactor. We have also said that there are many countries with nuclear power programmes which do not have associated nuclear weapons development. But is it possible to be *certain* that countries with such programmes will not use plutonium from civil reactors for military purposes?

The answer at present has to be 'no'. Unless the reprocessing of fuel from power reactors can be put under *effective* international control, the possibility of diverting plutonium for weapons use must always exist. Indeed, it is the reprocessing that provides the essential link between reactors and weapons. However, to ban all reprocessing (even were this possible) would be to exclude the development of breeder reactors (Section 3.4), and hence limit the ultimate energy potential of the world's uranium resources.

Taking this view a little further, an essay issued in 1991 by the British Nuclear Forum (a trade organization representing different sections of the nuclear industry) said:

> *...just as there is no technical fix for nuclear weapons proliferation—for example, closing all of the world's civil nuclear facilities would not guarantee any reduction in the threat—so it is agreed by the international community that safeguards are no substitute for removing the desire to develop nuclear weapons. Reduction in international tension is the most important requirement for future developments in the field. The Non-proliferation Treaty, in bringing together so many nations in this spirit, as well as making it possible to detect diversion of nuclear materials, has played an important role in allowing the right atmosphere to develop. But on its own it can never guarantee that nuclear weapons technology will not spread.*

Finally, we have been talking about possible links between civil nuclear power and nuclear weapons proliferation. However, are there any other ways in which the spread of nuclear weapons could increase? One of the greatest threats would appear to come from the events that have followed the collapse of communism in the former USSR. There is much concern about the fate of the stockpile of weapons-grade fissile material in the former Soviet republics, and about the future activities of their nuclear weapons experts in this relatively anarchic situation.

In conclusion the central question that remains, and one that you should address, is *would the cessation of the use of nuclear power world wide alter the risks of nuclear weapons proliferation?* This is, predictably, one of the open questions with which we are going to leave you.

Summary of Section 8.4

1   Nuclear weapons proliferation is supposedly controlled and prevented by the Non-proliferation Treaty. This gives the IAEA the power to inspect nuclear installations.

2   The case of Iraq suggests that without the full cooperation of signatories, it is very difficult, under normal circumstances, to make the Treaty fully effective.

3   Proper international control and inspection of both reactors and reprocessing facilities is one way of weakening possible links between nuclear power and nuclear weapons.

9 Energy and the future

Although nuclear fission is now firmly established as a source of electricity, and was used to make 16% of the world's electricity in 1990, it provides only about 6% of the *total* energy used. Another 6% of the world's energy needs was provided by hydro-electricity, but the majority (88%) came from the fossil fuels.

The depletion of the fossil fuel resources, the possible environmental effects of their use, and the increasing demand for energy have all given an impetus to the search for new energy supplies. In this book, we have until now been concerned with nuclear fission, but in this final chapter we shall look briefly at the progress that has been made towards harnessing the other nuclear energy source, fusion. We shall then look at the overall UK energy scene, in order for you to be able to put what you have learnt about nuclear power into perspective nationally.

9.1 Energy from fusion

There is a number of fusion reactions which have been considered as power sources. One is the reaction between two deuterium nuclei, ^2_1H, where two reactions are possible:

$$^2_1\text{H} + {}^2_1\text{H} \longrightarrow {}^3_2\text{He} + {}^1_0\text{n} + 3.2\,\text{MeV} \qquad (9.1)$$

or

$$^2_1\text{H} + {}^2_1\text{H} \longrightarrow {}^3_1\text{H} + {}^1_1\text{H} + 4.0\,\text{MeV} \qquad (9.2)$$

However, the reaction that is most likely to become a viable energy source, because it generates much more energy, is that between deuterium and tritium, ^3_1H (this is the same reaction as that which takes place in a fusion nuclear weapon—Section 8.1):

$$^2_1\text{H} + {}^3_1\text{H} \longrightarrow {}^4_2\text{He} + {}^1_0\text{n} + 17.6\,\text{MeV} \qquad (9.3)$$

▷ What are the reaction products in Equation 9.3?

▶ A helium-4 nucleus (an α-particle) and a neutron.

▷ In each of the fusion reactions in Equations 9.1 to 9.3, energy is produced. Could you have predicted from Figure 2.2, that fusing light elements in this way was likely to lead to a release of energy?

▶ Figure 2.2 shows that the average binding energy per nucleon tends to increase with increasing mass number, A, up to a value of A of about 60. This means that if two low-mass nuclei fuse to form a new nucleus, the binding energy per nucleon of the new nucleus will be greater than that for either of the two nuclei forming it. Consequently energy will be released. (Compare this with the argument used for nuclear fission in Figure 2.7.)

The energy produced in Equation 9.3 is shared between the α-particle and the neutron as kinetic energy, the neutron acquiring 14 MeV.

Deuterium is a naturally occurring isotope, forming 0.015% of natural hydrogen. Tritium, on the other hand, does not occur naturally. It is produced in reactors that use light or heavy water as a moderator or coolant (Section 5.3), has a half-life of 12.3 years, and decays by β-particle emission.

▷ How much energy is released when fission occurs?

▶ About 200 MeV (Section 2.3.1).

Although the energy released when deuterium and tritium fuse (17.6 MeV) is less than in fission, the mass of the two fusing nuclei is much less than that of a uranium or plutonium nucleus. As a result, the energy released per mass of 'fuel' (the fusing or fissioning nuclei) is actually *greater* from the fusion of deuterium and tritium than from fission. Let us see how much more by looking at the energy released per nucleon involved.

The deuterium and tritium nuclei together contain a total of 5 nucleons. So the energy released per nucleon is (17.6 MeV/5), which is 3.5 MeV. In the case of fission of a ^{235}U nucleus, the energy released is 200 MeV (Section 2.3.1), so the energy released per nucleon is (200 MeV/235), which is 0.85 MeV. Thus, the energy released in the fusion reaction per nucleon is some four times greater than in fission.

This seems so attractive that there must be a problem. There is: it is making fusion happen!

▷ Why is it difficult to make two nuclei fuse?

▶ Nuclei are positively charged and positive charges repel each other. So, if we try to fuse deuterium and tritium nuclei, we have to provide the nuclei with enough energy to overcome the repulsive force between them.

There are several ways in which this can be achieved. In Chapter 8 we saw that in a nuclear fusion weapon a fission weapon is used to heat up and compress the deuterium and tritium to make them fuse.

▷ How does this make fusion possible?

▶ Heating up the deuterium and tritium gives the nuclei kinetic energy. At a sufficiently high temperature, this kinetic energy will be large enough to overcome the repulsion between the nuclei. The explosion also compresses the deuterium–tritium mixture, making it more likely that the nuclei will collide.

In trying to harness fusion for peaceful uses, the same idea can be used; a mixture of deuterium and tritium gas can be heated to very high temperatures (in excess of 10^8 °C) and 'squeezed' together using very intense magnetic fields. At these very high temperatures, the atoms split up into nuclei and electrons, which move about independently; this mixture is called a **plasma**. (The fact that high temperatures are involved in getting this nuclear reaction to occur is why it is often called a **thermonuclear reaction**.)

If enough collisions occur between the deuterium and tritium nuclei, more energy can be generated than is used to produce the plasma. The system will then become an energy source. The Sun is an example of a very successful fusion system, although its energy is not based on the deuterium–tritium fusion reaction.

9.1.1 Harnessing nuclear fusion

The use of nuclear fission to generate electricity was greeted with tremendous enthusiasm in the 1950s, as a headline from *The Times* supplement of 17 October 1956, commemorating the opening of the Calder Hall power station, shows:

'The second industrial revolution'   'Nuclear power the basis'

Two years after this, the possibility that nuclear fusion could *also* be harnessed was greeted with a similar fanfare, as can be seen from Extract 9.1, which is the report of a breakthrough on ZETA (Zero Energy Thermonuclear Assembly) at Harwell. Let us look at this through Activity 9.1.

Activity 9.1

Read Extract 9.1.

(a) The Harwell scientists claimed to have detected a fusion reaction in ZETA. What particle did they detect to enable them to substantiate their claim? Which of the reactions in Equations 9.1–9.3 would have been responsible?

(b) Why did the ZETA experiments suggest that an inexhaustible supply of energy might be obtained from the oceans?

Extract 9.1   From *The Manchester Guardian*, 25 January 1958.

PROGRESS TOWARDS H-POWER

Zeta unveiled: Sir John Cockcroft's hopes

FUEL FROM THE OCEANS

By our scientific correspondent

Design of a machine in which temperatures of 100,000,000 degrees will be created is about to begin in Harwell. In this machine it is hoped that thermo-nuclear energy from deuterium will be produced in quantities comparable with the power needed to operate the machine.

In the meantime, the experimental machine Zeta is being modified to reach temperatures in excess of the 5,000,000 degrees already attained, and should produce temperatures greater than 15,000,000 degrees "within a year".

These facts have been announced by Sir John Cockcroft as part of a release of information about thermo-nuclear research made by the Atomic Energy Authority and the United States Atomic Energy Commission last night. Sir John Cockcroft said that he thought it would take "twenty years plus" before thermo-nuclear reactions would be put to practical peaceful use.

An A.E.A. statement issued in London sounded the same note of cautious optimism. It said in part:

"Many major problems have still to be solved before its practical application can be seriously considered, and the work must be expected to remain in the research stage for many years yet. If it proves ultimately possible to construct a power station operating on the fusion of deuterium (heavy hydrogen), the oceans of the world will provide a virtually inexhaustible source of fuel."

Details of recent experiments with the Zeta machine at Harwell were also made public yesterday. Temperatures of five million degrees (Centigrade) are said to have been produced for periods of about four-thousandths of a second in an electrical discharge in deuterium. During the same period bursts of about a million neutrons have been observed to be released from the machine.

"90% sure"

Officially it is not claimed that these neutrons are produced by true thermo-nuclear reactions within the discharge tube. However, Sir John Cockcroft is "90% sure" that they have come from the kind of nuclear process which keeps the sun hot and which makes hydrogen bombs possible. "It will be very surprising if they are not thermo-nuclear neutrons," he said.

Sir John Cockcroft said yesterday that the immediate object of research in this field would be to achieve higher temperatures in gas discharges and to maintain these for greater lengths of time. He mentioned 100 million degrees as the temperature at which the power from thermo-nuclear reactions might "break even" with that needed to keep machines going. (The temperature at the centre of the sun, where similar reactions take place, is about 10 million degrees centigrade). Sir John said that these temperatures would have to be maintained for times not much less than a second. The possibility of converting the power directly into electricity was exciting.

Most of the information released is published in the current issue of the scientific journal

"Nature" (which yesterday was being sold as if it were an evening paper outside the gates at Harwell). It includes an article from Harwell describing work with Zeta, another from the research laboratories of A.E.I., Ltd., at Aldermaston, describing work with a smaller machine called "Sceptre III", and four articles by scientists of the Atomic Energy Commission.

" Why shouldn't I grow up into a lazy good-for-nothing—I'll be living in the age of Zeta, won't I ? "

YEARS OF PROGRESS

The following are some of the significant dates in research on thermonuclear reactions:

1932—First artificial transmutation of pairs of deuterium nuclei into helium at the Cavendish Laboratory.

1935–7—Description of the thermo-nuclear reactions which, it can be inferred, are responsible for the production of energy in stars.

1945–7—Research started by Thonemann at the Clarendon Laboratory, Oxford, and by Cousins and Ware at Imperial College.

1950—Thermo-nuclear research declared secret: Oxford group transferred to Harwell and Imperial College group to A.E.I. research laboratories, Aldermaston.

1955 (August)—Public acknowledgement by the United Kingdom, the United States, and U.S.S.R. that thermo-nuclear research was in progress.

1956—In May Academician Kurchatov, speaking at Harwell, described an unsuccessful thermo-nuclear experiment.

1957—In June the United States announced a plan to build a big thermo-nuclear machine called "Stellerator."

1957—August 17, Zeta commissioned; August 30, Zeta produced neutrons; October 25, meeting between British and American groups at Princeton; December, "Perhapsotron" and "Sceptre III" produce neutrons.

The slight note of caution voiced by Sir John Cockcroft (the Chairman of the UKAEA) in this article (he was '90% sure'), but spelt out in more detail in other articles and in the associated press release, concerned whether or not the neutrons observed came from fusion occurring in the plasma in the way we described earlier. Thus, the press release from the UKAEA on 24 January 1958 said:

> *The source of the observed neutrons has not yet been established. There are good reasons to think that they come from thermo-nuclear reactions, but they could also come from other reactions such as collision of deuterons [deuterium ions] with the walls of the vessel, or from bombardment of stationary ions by deuterons accelerated by internal electric fields produced in some form of unstable discharge.*

The feeling of euphoria was short lived. The caution expressed was justified; it turned out that the neutrons had *not* originated from fusion reactions between deuterium ions in the heated plasma. After this disappointment, it became apparent that the scale and cost of nuclear fusion research made international collaboration the most fruitful way forward. In Europe this resulted in the JET project, the Joint European Torus, for which site construction started in 1979. (A *torus* is the name given to the doughnut-shaped containment vessel which is used, as it also was in ZETA.)

Fourteen nations (all the European Community together with Sweden and Switzerland) have been involved in building and testing the JET assembly at Culham, near Oxford (Figure 9.1). After nearly a decade of operation, during which time the performance and characteristics of the system were studied using deuterium only, some tritium was added to the deuterium fuel. Activity 9.2 is concerned with the progress that has been made in this project.

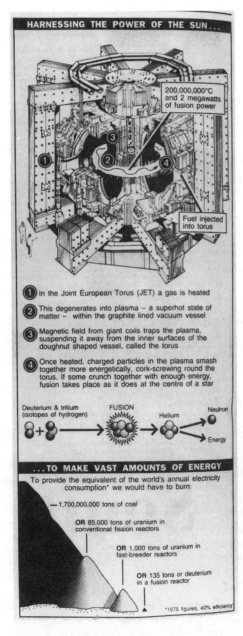

Figure 9.1 The JET assembly (from *Daily Telegraph*, 11 November 1991).

Activity 9.2

Extract 9.2 is the JET project press release describing the first extraction of significant amounts of power from controlled nuclear fusion reactions. Extract 9.3 is a newspaper article covering the same event. Read both extracts and then answer the following questions.

(a) Write down the equation for the fusion reaction that was involved. In what two important respects does it differ from the reaction that was thought to have been observed in ZETA (Activity 9.1)?

(b) Assuming that the press release is a correct description of the JET experiment, what, if any, inaccuracies can you detect in the newspaper report?

(c) Newspaper and television reports are sometimes accused of making scientific work and scientific phenomena sound more sensational than they really are. Are the errors that you detected in part (b) of this nature?

Extract 9.2   JET press release, 9 November 1991.

Jet achieves fusion power

At 7.44 pm today, Saturday 9th November 1991, between 1 500 000 and 2 000 000 watts of power from nuclear fusion reactions were generated at the JET (Joint European Torus) collaborative European Community project based at Abingdon, Oxfordshire, UK.

The Director of JET, Dr Paul-Henri Rebut announcing the successful experiment said "this is the first time that a significant amount of power has been obtained from controlled nuclear fusion reactions. It is clearly a major step forward in the development of fusion as a new source of energy."

Today's experiment was the first occasion in which the correct fusion fuels, deuterium and tritium, have been used in any magnetic confinement fusion experiment. Previously the experimental performance had been such as to justify only the use of deuterium fuel in which the fusion reaction rate is much slower. Since the start of operation in 1983, JET's performance has progressively approached reactor conditions. The planned operation with the correct mix of the reactor fuels—a 50/50 mixture of deuterium and tritium—is being approached in a stepwise manner, with the first step being the present experiments at low tritium concentrations and concluding at full power operation with 50% tritium.

In today's experiment the deuterium and tritium gas was heated to temperatures of around 200 million degrees Celsius—nearly more than 10 times hotter than the temperature in the centre of the sun. The peak fusion power generated reached almost 2 000 000 watts (2 MW). In a pulse lasting for two seconds and giving a total energy release equivalent to a megawatt for two seconds. At lower power in deuterium JET has already maintained stable conditions in the apparatus for periods up to 1 minute.

JET is a collaborative venture involving all countries of the European Community, together with Switzerland and Sweden. As the world's largest fusion device, JET has achieved separately all individual parameters required in a reactor. The data obtained from JET has laid a firm foundation for the proposed experimental reactor ITER (International Thermonuclear Experimental Reactor), which is planned to be carried out as a world-wide collaboration involving the United States, Japan, Soviet Union and the European Community.

"The hard work and dedication of all the JET staff over many years, together with the support of the European Nations who are members of the Joint Undertaking have today been rewarded by this achievement" said Dr. Rebut. "These experiments are a significant milestone and clearly confirm Europe's leading position in fusion research. This demonstration fully confirms that with the additional information from the planned JET programme up to 1996 we will be able to design the experimental fusion reactor ITER capable of generating more than 1 000 megawatts of thermal power".

Extract 9.3 From *The Independent on Sunday*, 10 November 1991.

Nuclear fusion a giant step closer

By Richard Woodman and Tom Wilkie

Scientists took a historic leap forward yesterday in harnessing nuclear fusion—the way in which the Sun produces its power—and converting it to a controllable source of energy.

Nuclear fusion is the opposite of nuclear fission, which is what happens inside a nuclear power station. Instead of unchaining vast quantities of energy by splitting the heaviest atoms—such as uranium and plutonium—nuclear fusion obtains more energy by fusing atoms of the lightest element, hydrogen, converting it to helium and throwing off quantities of neutrons at high speed.

Nuclear fusion has previously been achieved for only a second or two; yesterday's experiment at the Joint European Torus (JET) reactor at Culham in Oxfordshire managed it for two minutes, bringing close the prospect of an apparatus capable of producing usable quantities of power.

JET scientists have been using a form of "heavy" hydrogen known as deuterium to produce nuclear fusion. Yesterday, they added tritium—yet another form of hydrogen—to make "real" fusion reactor fuel, which resulted in a much more energetic nuclear reaction, and produced the equivalent of a million watts of power inside the "Torus", a reactor chamber three times taller than a double-decker bus.

If a way can be found to sustain fusion for sufficiently long periods, mankind can look forward to the prospect of virtually limitless—and relatively clean and safe—supplies of energy. The fuels needed for fusion are plentiful; deuterium is extracted from water while tritium can be made inside the reactor.

Ten grams of deuterium and 15 grams of tritium would meet the lifetime electricity needs of an average person in an industrialised country.

However, the prospect of commercial fusion reactors still appears several decades away.

John Maple, a JET spokesman, said: "It is the first time that anyone has produced any substantial amount of fusion power in a controlled fusion experiment as opposed to a bomb."

Unlike the excitement caused last year when hopes of achieving cold fusion inside a simple test-tube were first raised, then dashed, yesterday's experiment involved temperatures of up to 300 million degrees centigrade, 20 times hotter than the sun.

Britain and 13 other countries involved in the JET project, which costs £75m a year, now enjoy a world lead in trying to harness the power of the Sun on Earth. They are significantly ahead of their American rivals at the Tokamak Fusion Test Reactor at Princeton and ahead of the Japanese JT-60 fusion reactor.

Only a tiny amount of tritium—about 0.2 gram—was used in yesterday's experiment. The concentration used was only 14 per cent, compared with 86 per cent deuterium. More energy would have been produced had the scientists gone for the ultimate goal of a 50–50 mix.

Mr Maple said: "Fusion is extremely safe. Anything you do stops the reaction. Even if something did go wrong there is so little fuel it would not create an emergency situation that would require evacuation of the population around the reactor."

Unlike coal and oil-fired power stations, the process does not cause any atmospheric pollution or global warming.

In this case there is no doubt that the fusion of deuterium and tritium took place under the conditions that could lead to fusion becoming a power source. However, this test does not prove that a fusion reactor capable of producing power continuously can be built. Indeed, during the test described, the power produced was about 1 MW, and the power input was 700 MW. Nevertheless, assuming that the progression towards a true fusion reactor takes place as those involved hope, it is worth noting the time-scale being mentioned—several decades according to Extract 9.3 (other newspaper reports say 40–50 years). This means that any benefits from fusion are unlikely to be realized for several generations.

In view of the uncertainties involved, it would be very unwise to make plans based on the availability of fusion power. On the other hand, it might be equally unwise to close the door to its continued development. On a global scale, the costs involved are small, and it would seem sensible to explore as many avenues as possible to decrease our dependence on fossil fuels. If research and development funds for energy are limited, it then becomes necessary to try to determine which of the many activities competing for funds—the renewable sources, energy conservation, fusion and fission breeder reactors, etc.—are *likely* to give the greatest return on investment. In the UK the investment choice made in recent years was indicated in Figure 7.5. It is in making the judgement of how to allocate investment funds that the heart of the problem lies, because, apart from the intrinsic difficulties, these judgements are influenced by pressure groups and political climates.

From *The Guardian*, 15 November 1991.

Summary of Section 9.1

1   Of the fusion reactions that have been considered as power sources, that between deuterium and tritium produces most energy.

2   Because the positively charged nuclei repel one another, controlled fusion reactions must be conducted at very high temperatures while the nuclei are contained and squeezed together, for example by intense magnetic fields.

3   After more than 50 years of research, controlled fusion has been achieved for very brief periods of time, but it has not yet been harnessed to produce electricity. Those in the field still speak of a fusion reactor for producing electrical power being decades away.

9.2   The UK energy scene

This book has been concerned with nuclear power, and in concentrating on one energy source in this way, you have not had the chance to obtain an overall perspective on energy production and use. Yet if you are to reach an informed opinion on nuclear power, you need to know how it fits into the overall energy scene. In this short section we shall therefore very briefly consider the provision and use of energy in the UK, looking carefully at some pertinent data.

9.1

Box 9.1   The energy units used

Throughout this book we have been using either the electron volt (eV) or the joule (J) to measure energy, and the watt (W) for power—the rate of energy production or use. However, in the energy industry the units are varied—and descriptive! For example, oil may be measured by mass (tonnes) or by volume (barrels). In addition, some old energy units are still in use, like the *therm*. When energy sources are being compared they are often converted into the *equivalent* measures of oil or coal. Where it is important for comparison purposes, we shall convert energy into SI units; the average conversion factors are given in Table 9.1. Nevertheless, you should get used to looking at energy data in different units, so we shall not convert all the data presented.

▷ What is the energy in a barrel of oil expressed in tonnes of coal equivalent?

▶ A barrel of oil has an energy content of 5.7×10^9 J. A tonne of coal has an energy content of 24×10^9 J. So, a barrel of oil has an energy equivalent to $(5.7 \times 10^9/24 \times 10^9)$ t of coal = 0.24 t of coal.

Table 9.1 Average values for the energy conversion factors for fossil fuels. We say *average* values because the energy content can vary from one sample of coal or oil to another.

Resource	Unit of amount	Energy content/joules
coal	tonne	24×10^9
oil	barrel	5.7×10^9
	tonne	42×10^9
gas	therm	105×10^6
	m³	38×10^6
electricity	kW h	3.6×10^6 ■

Question 9.1 What is the energy content of a tonne of oil, expressed in terms of the equivalent mass of coal?

9.2.1 UK energy supply and demand

The energy supply system is a dynamic one, both nationally and internationally, and changes can sometimes occur quite rapidly, that is, over ten years or less. There can be even more rapid changes arising from national or international problems—for example a strike, or political action by a supplier. Let us see if there are signs of these and other influences in the pattern of UK energy consumption during the past 30 years.

Total UK energy consumption grew from 160 million tonnes of oil equivalent in 1960 to about 210 million tonnes of oil equivalent in 1990. This is an increase of about 30%. This increase, however, was modest in comparison with the increase in world energy consumption, which nearly trebled in the same period. The implication of this increase in world energy use is that it will place increasing pressure on energy supplies; it will also exacerbate the problem of fossil fuel emissions such as CO_2.

Let us now see how the purposes for which energy is used in the UK have changed over the years. The pie charts in Figure 9.2 divide energy consumption into four important categories of end user.

▷ What two big changes occurred between 1960 and 1990?

▶ There was a big increase in energy use for transport (from 17% to 33%), whereas use by industry fell sharply (from 42% to 27%).

We next turn to the ways in which the *type of fuel* used by consumers changed over the same period. This is shown by the pie charts in Figure 9.3.

Figure 9.3 reveals the substantial increase in the consumption of energy as oil (petroleum). This is consistent with Figure 9.2, in that much of the increased petroleum consumption will have been needed for the increased use of energy for transport,

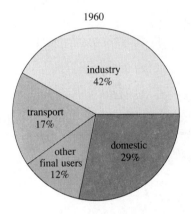

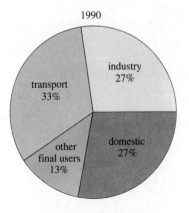

Figure 9.2 Energy consumption in the UK in four different end user categories for the years 1960 and 1990.

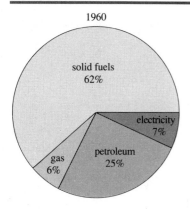

1960

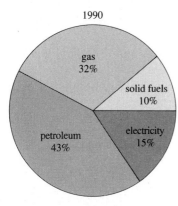

1990

Figure 9.3 Energy consumption in the UK in 1960 and in 1990 according to the type of fuel. The data ignore the energy used/wasted in the *generation* of electricity.

which requires fuels such as petrol, diesel oil and kerosene. Figure 9.3, however, reveals two other changes that took place between 1960 and 1990, which are more dramatic than the increased consumption of petroleum.

▷   What are they?

▶   There was a dramatic fall in the use of solid fuels such as coal, and a big rise in the consumption of gas.

In this book, however, our main concern has been with electricity production because that is where nuclear power makes its contribution. Figure 9.3 shows that alongside the other changes that we have mentioned there was a substantial increase in the percentage of energy consumed as electricity (from 7% to 15%). Moreover, because, as noted earlier, total energy consumption increased at the same time, the *amount* of energy consumed as electricity grew even more than these percentages would suggest. To be specific, in 1960, electricity consumption was about 10 million tonnes of oil equivalent; in 1990, it was about 25 million tonnes: the increase is about 150%. Nevertheless, despite this great increase in importance, electricity still accounts for only a small proportion of *total* energy consumed, as the 1990 figure of 15% shows.

Figure 9.4 breaks down total electricity production in the period 1960–90 into nuclear and non-nuclear (conventional thermal, that is, fossil fuel) contributions. The total amount of electricity generated has risen from about 120 TW h (1 TW h = 10^{12} Wh) to about 300 TW h. At the start of this period, the contribution from nuclear electricity was negligible; only the small plutonium-producing reactors at Calder Hall and Chapel Cross were operating. By 1990, the contribution from nuclear electricity was about 50 TW h or 17%. Thus, during this period, nuclear power made a substantial contribution to the growth of electricity generation: it made up nearly 30% of the 180 TW h increase.

Figure 9.4 Nuclear and non-nuclear contributions to electricity production in the UK from 1960 to 1990.

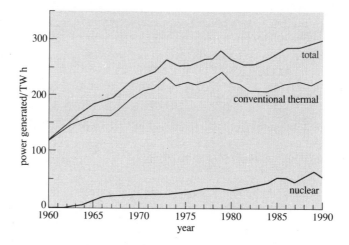

At the time of writing (1993) nuclear power in Britain was still provided by more than 20 Magnox reactors and 14 AGRs, though a PWR is being constructed at Sizewell. Figure 9.5 shows the location of the various types of British nuclear reactors and other nuclear installations.

Figure 9.6 shows the total energy used in electricity generation, separated into coal, oil and nuclear. This includes the large amount of energy wasted as heat during the generating process (Section 3). You can see that prior to 1973, increasing amounts of

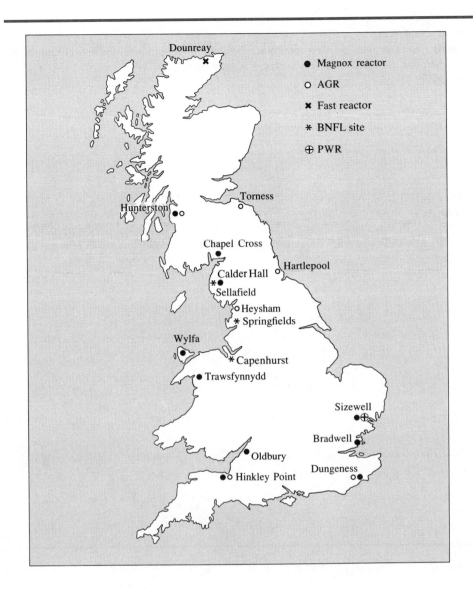

Figure 9.5 Map showing the location of British nuclear reactors and BNFL's fuel fabrication and processing installations.

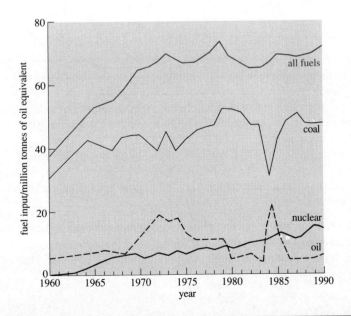

Figure 9.6 Fuel input into electricity generation in the UK for the period 1960–90.

oil were devoted to electricity production, but thereafter, the price rises occasioned by the Yom Kippur War of that year brought about a decline. There are also irregularities which are best discussed by noting the sharp dips in the plot for coal in 1972, 1974 and 1984.

▷   To what do you attribute these?

▶   They mark the miners' strikes of 1972, 1974 and 1984–5.

The dips in coal input in Figure 9.6 (1972, 1974 and 1984–5) are matched by peaks in oil input, showing that attempts were made in each coal strike to compensate for the loss of coal with an increased use of oil. Nuclear power could not respond so flexibly, but there are still detectable peaks in the nuclear plot for those periods, and nuclear generation, of course, made its steady contribution to the base load.

The UK energy resources

By comparison with many countries, and particularly other industrialized European ones like France, the UK is extremely well endowed with energy sources. It has very large resources of coal, and is currently self-sufficient in oil and natural gas, though it has little uranium. Indeed, as a result of oil and gas production from the North Sea, the UK became a net energy exporter in the 1980s (Figure 9.7); this was the first time this had happened since the start of the Second World War.

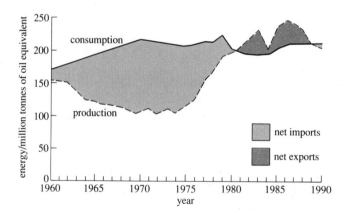

Figure 9.7   UK energy production and consumption.

In addition to fossil fuels, the UK has large sections of coastline which would be suitable for harnessing wave power, several off-shore and coastal areas where wind power could be exploited, and there are some potential sites for tidal power stations, notably the Severn estuary (Figure 9.8). Continuous sunshine is one of the natural energy sources the UK lacks; it also has little land that could be used to exploit biomass energy sources (e.g. wood and plant matter, which can provide a valuable alternative to oil), and few unexploited sites for hydroelectricity.

However, the UK has sufficient resources to try to balance the conflicts between safety, cost, sustainability and environmental impact. Making a *proper* assessment of all these factors for the different energy sources is the starting point for striking such a balance—and an extremely difficult one.

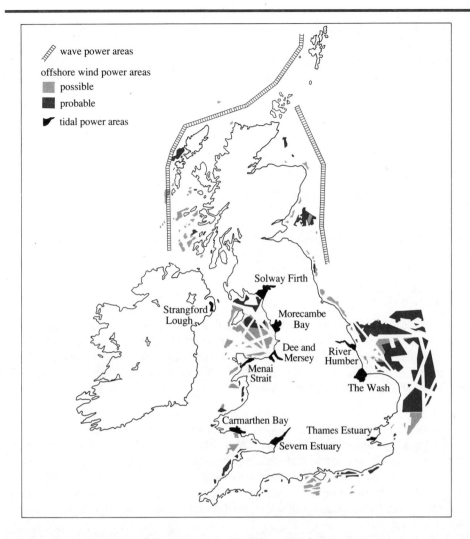

Figure 9.8 Potential sites for electricity generation by tidal, wave and offshore wind power in the UK.

Summary of Section 9.2

1   Between 1960 and 1990, UK energy consumption increased by 30%, while world energy consumption trebled. In the UK there was a big shift away from solid fuels to the use of natural gas and oil, the latter being encouraged by increased energy use for transportation.

2   Throughout this period, electricity consumption accounted for less than 20% of total energy used, but the proportion increased very substantially by about 150%.

3   In 1990, nuclear power supplied 17% of UK electricity generation, and it accounted for nearly 30% of the increase in electricity generation that took place between 1960 and 1990.

4   The ability of nuclear power to act as an energy source is limited by the fact that it is currently used only to supply electricity. It adds variety to the fuels used for electricity production, and thus contributes to security of supply.

5   The UK is very well endowed with both fossil fuels and renewable energy sources.

Question 9.2   Section 9.2.1 points out that between 1960 and 1990, electricity consumption in the UK increased from about 10 million to 25 million tonnes of oil equivalent. Figure 9.6 shows the fuel *input* to electricity generation in these years. Estimate the average efficiency of electricity generation in 1960, and in 1990, and comment on any difference.

9.3 In conclusion

As in so many human activities, the provision of energy requires a compromise between a number of factors. In this book you have studied the use of nuclear fission to produce electricity and have found that it raises a number of issues such as cost, safety and waste disposal. You have also considered the relationship between nuclear power and the proliferation of nuclear weapons.

In Chapter 1 we said that the aim of this book is 'to present the issues involved and the background to them: it will then be up to *you* to make any judgement for yourself'. To fulfil this task, it was necessary to 'provide information on the scientific and technological background to nuclear power'.

How can you judge if these aims have been achieved in your case? You could start by referring back to your answers to Activity 1.1, and considering whether any of your responses have changed since studying this book. You should also find that you are better able to follow articles on nuclear power in the media. Some articles on the subject have been collected together in the offprints for this book, and you could read these now.

When you read these articles, you should find that you are able to *understand* them, perhaps not every detail at first reading, but well enough to grasp what the writer is trying to say. The issues will be familiar to you, and you now have the necessary background to appreciate the technical points being made, and hence to determine the soundness of the arguments advanced.

You now also know that many issues are not black and white; there are uncertainties in the technical and economic arguments, and ample scope for differences of opinion about what constitute acceptable goals. All this knowledge is an essential prerequisite for having an informed opinion. However, because there *are* uncertainties, issues are rarely resolved once and for all. Hence, your views on an issue could change in the light of new information or considerations—indeed, they *should* change. To stick to opinions through thick and thin, however well founded they were originally, is inconsistent with the aims of this book.

Appendix 1 Relative atomic masses and atomic numbers of the elements

Element	Symbol	Atomic number	Approximate relative atomic mass	Element	Symbol	Atomic number	Approximate relative atomic mass
actinium	Ac	89	227	hydrogen	H	1	1.01
aluminium	Al	13	26.98	indium	In	49	114.82
americium	Am	95	243	iodine	I	53	126.90
antimony	Sb	51	121.75	iridium	Ir	77	192.2
argon	Ar	18	39.95	iron	Fe	26	55.85
arsenic	As	33	74.92	krypton	Kr	36	83.80
astatine	At	85	210	lanthanum	La	57	138.91
barium	Ba	56	137.34	lawrencium	Lr	103	257
berkelium	Bk	97	247	lead	Pb	82	207.2
beryllium	Be	4	9.01	lithium	Li	3	6
bismuth	Bi	83	208.98	lutetium	Lu	71	174.97
boron	B	5	10.81	magnesium	Mg	12	24.31
bromine	Br	35	79.91	manganese	Mn	25	54.94
cadmium	Cd	48	112.40	mendelevium	Md	101	256
caesium	Cs	55	132.91	mercury	Hg	80	200.59
calcium	Ca	20	40.08	molybdenum	Mo	42	95.94
californium	Cf	98	251	neodymium	Nd	60	144.24
carbon	C	6	12.01	neon	Ne	10	20.17
cerium	Ce	58	140.12	neptunium	Np	93	237.05
chlorine	Cl	17	35.45	nickel	Ni	28	58.71
chromium	Cr	24	52.00	niobium	Nb	41	92.91
cobalt	Co	27	58.93	nitrogen	N	7	14.01
copper	Cu	29	63.54	nobelium	No	102	254
curium	Cm	96	247	osmium	Os	76	190.2
dysprosium	Dy	66	162.50	oxygen	O	8	16.00
einsteinium	Es	99	254	palladium	Pd	46	106.4
erbium	Er	68	167.26	phosphorus	P	15	30.97
europium	Eu	63	151.96	platinum	Pt	78	195.09
fermium	Fm	100	253	plutonium	Pu	94	242
fluorine	F	9	19.00	polonium	Po	84	210
francium	Fr	87	223	potassium	K	19	39.1
gadolinium	Gd	64	157.25	praseodymium	Pr	59	140.91
gallium	Ga	31	69.72	promethium	Pm	61	147
germanium	Ge	32	72.59	protactinium	Pa	91	231.04
gold	Au	79	196.97	radium	Ra	88	226.03
hafnium	Hf	72	178.49	radon	Rn	86	222
helium	He	2	4.00	rhenium	Re	75	186.2
holmium	Ho	67	164.93	rhodium	Rh	45	102.91

Element	Symbol	Atomic number	Approximate relative atomic mass	Element	Symbol	Atomic number	Approximate relative atomic mass
rubidium	Rb	37	85.47	thallium	Tl	81	204.37
ruthenium	Ru	44	101.07	thorium	Th	90	232.04
samarium	Sm	62	150.35	thulium	Tm	69	168.93
scandium	Sc	21	44.96	tin	Sn	50	118.69
selenium	Se	34	78.96	titanium	Ti	22	47.90
silicon	Si	14	28.09	tungsten	W	74	183.85
silver	Ag	47	107.89	uranium	U	92	238.03
sodium	Na	11	22.99	vanadium	V	23	50.94
strontium	Sr	38	87.62	xenon	Xe	54	131.30
sulphur	S	16	32.06	ytterbium	Yb	70	173.04
tantalum	Ta	73	180.95	yttrium	Y	39	88.91
technetium	Tc	43	98.91	zinc	Zn	30	65.37
tellurium	Te	52	127.60	zirconium	Zr	40	91.22
terbium	Tb	65	158.92				

Further reading

Beckmann, P. (1976) *The Health Hazards of NOT Going Nuclear*, Golem Press, Colorado.
Adopts the readable, punchy style normally associated with some of the anti-nuclear publications to paint a pro-nuclear picture, by concentrating on the risks from coal in the USA.

Berkhout, F. (1991) *Radioactive Waste: Politics and Technology*, Routledge.
Using Germany, Sweden and the UK as examples, the technical and political issues surrounding radioactive waste disposal are set out in a readable way. Supports Chapter 5.

Blowers, A., Lowry, D., and Solomon, B. (1991) *The International Politics of Nuclear Waste*, Macmillan.
A comprehensive and detailed coverage of the political problems involved in radioactive waste disposal, and of the opposition to it. Supports Chapter 5.

British Nuclear Energy Society (1987) *Chernobyl, a Technical Appraisal*, British Nuclear Energy Society.
The report of a conference held soon after the accident. It contains useful data but is technical and only accessible with effort. Supports the part of Chapter 6 which is concerned with Chernobyl.

Cameron, I. R. (1982) *Nuclear Fission Reactors*, Plenum Press.
Although a specialist book going far beyond the requirements of *Science Matters*, this is clearly written (and hence accessible with effort) and contains much useful data. Supports Chapter 3 in particular.

Chapman, N. A., and McKinley, I. G. (1987) *The Geological Disposal of Nuclear Waste*, Wiley.
A specialist book but accessible to the non-expert; the basic ideas of radioactive waste disposal are clearly explained. Supports Chapter 5.

Davies, P. (1988) *Magnox: the Reckoning*, Friends of the Earth.
A concise summary of the objections to the Magnox reactor programme.

Fremlin, J. H. (1989) *Power Production: What are the Risks*, 2nd edn, Hilger.
An extremely readable and carefully argued book, which sets out the case for nuclear power by careful analysis of the counter arguments.

King, P. (1990) *Nuclear Power: the Facts and the Debate*, Quiller Press.
A very basic, simply written but wide-ranging book, with many illustrations.

Nero, A. V. (1979) *A Guidebook to Nuclear Reactors*, University of California Press.
Although going far beyond the requirements of *Science Matters*, this book has chapters on many of the topics discussed in NP, together with a host of data.

Patterson, W. C. (1986) *Nuclear Power*, 2nd edn with postscript, Penguin.
An updated version of one of the earlier and more comprehensive texts looking critically at nuclear power. Very accessible.

Sumner, D. (1990) *Radiation Risks: an Evaluation*, 3rd edn, Tarragon Press.
A short and very clearly written book, which contains much useful data and concludes with a discussion of the Chernobyl accident. It is also of particular relevance to Chapter 4.

Skills

In this section, we list skills that have been explicitly taught and/or revised in this book. You should find that most of them are special instances of the general skills categories listed in the *Course Study Guide*. As usual, some of them (6 and 7) are rooted in the content of the book, some have been extensively practised in earlier books (for example 1, 3 and 5), but others are relatively new (for example 8 and 9).

1   Interpret and manipulate data presented in the form of text, tables, graphs, bar charts, pie charts and diagrams. (Questions 2.7, 3.1, 5.2–5.5, 6.1, 7.1–7.3 and 9.2; Activities 3.1, 3.2, 4.1–4.3, 5.2 and 7.2)

2   Manipulate mathematical formulae to obtain the value of a variable in the formula when the other variables take fixed and given values. (Questions 2.2, 2.6, 3.2 and 5.1; Activity 5.2)

3   Convert scientific quantities from one set of units to another. (Question 9.1)

4   From a section of teaching text you have studied, extract information relevant to a particular problem. (Questions 2.3, 2.8, 2.9 and 4.1; Activity 6.1)

5   Extract from an article, parts of which you may find unintelligible, information that is relevant to a particular question, and by integrating that information with what you already know, give an answer to the question in your own words. (Activities 4.4, 4.5, 5.2, 6.3, 6.4, 6.5, 7.1, 7.3, 8.1, 8.2 and 9.1)

6   Show familiarity with the notations used to identify isotopes, and use them to write balanced equations for nuclear processes. (Questions 2.1, 2.4, 2.5 and 2.7; Activity 2.2)

7   Construct simple event trees, and given the probabilities of the individual incidents of which they are composed, calculate the probabilities of the different probable outcomes. (Activity 6.6)

8   Use a scientific model to draw conclusions, showing at the same time, an awareness of the model's deficiencies. (Activity 2.1)

9   Transform arguments expressed in verbal reasoning into mathematical equations, and use the equations to obtain useful results. (Question 3.3; Activity 5.1)

10 Use information obtained from one source to comment on, or criticize views expressed in, another. (Question 8.1; Activities 3.2, 4.5, 8.2 and 9.2)

11 Formulate a personal opinion on a scientific issue. (Activities 1.1, 6.2–6.4, and 8.1)

Answers to questions

Question 2.1

The mass number, A, is the sum of the number of protons and neutrons. For hydrogen this would be $(1 + 0)$, for sodium $(11 + 11)$, and for lead $(82 + 126)$. Since the atomic number, Z, is the number of protons, the symbolic representations for these isotopes in our first notation would therefore be ^1_1H, $^{22}_{11}\text{Na}$ and $^{208}_{82}\text{Pb}$. In the second notation, the isotopes would be written ^1H or hydrogen-1, ^{22}Na or sodium-22, and ^{208}Pb or lead-208.

Note that ^1H does not have a neutron in its nucleus. In this respect, it is unique; all other isotopes have at least one neutron.

Question 2.2

There are 8 protons and 8 neutrons in oxygen-16. Using the data in the text, their total mass is $(8 \times 1.672\,62 \times 10^{-27}\,\text{kg})$ for the protons and $(8 \times 1.674\,93 \times 10^{-27}\,\text{kg})$ for the neutrons. The total mass of the individual protons and neutrons is therefore:

$$(13.380\,96 + 13.399\,44) \times 10^{-27}\,\text{kg} = 26.780\,40 \times 10^{-27}\,\text{kg}$$

The mass of the oxygen-16 nucleus is $26.552\,97 \times 10^{-27}\,\text{kg}$. The reduction in mass when the oxygen-16 nucleus is formed from individual neutrons and protons is thus

$$(26.780\,40 - 26.552\,97) \times 10^{-27}\,\text{kg} = 0.227\,43 \times 10^{-27}\,\text{kg}$$

We now find the energy equivalence of this mass change, which is:

$$0.227\,43 \times 10^{-27} \times (2.998 \times 10^8)^2\,\text{J} = 2.044 \times 10^{-11}\,\text{J}$$

The energy in eV is then $2.044 \times 10^{-11} \times 1/(1.602 \times 10^{-19})$

$$= 127.6 \times 10^6\,\text{eV} = 127.6\,\text{MeV}$$

The *total* binding energy for oxygen-16 is therefore 127.6 MeV. There are 16 neutrons and protons, so the binding energy per nucleon in oxygen-16 is therefore $(127.6/16)\,\text{MeV} \approx 8.0\,\text{MeV}$.

Question 2.3

(a) *True*. With one exception (^1_0H) all nuclei contain protons and neutrons; uncharged atoms are surrounded by the same number of electrons as there are protons in the nucleus.

(b) *False*. The volume of the nucleus of an atom is a *very* small proportion of the total volume of the atom.

(c) *True*. Mass and energy can be considered to be equivalent.

(d) *False*. As the number of protons, Z, increases, so does the proportion of neutrons needed to keep the nucleus stable, and hence the ratio of neutrons to protons increases.

(e) *False*. The nuclear binding energy is the energy released when the nucleus is *formed* from its constituent neutrons and protons.

(f) *True*. Energy equal to the binding energy is released when a nucleon is added to a nucleus.

Question 2.4

(i) The mass number is unchanged but the atomic number has decreased by one; $^{15}_{8}\text{O}$ therefore decays by positron emission to $^{15}_{7}\text{N}$.

(ii) The mass number is reduced by four, and the atomic number by two, so a helium nucleus has been ejected. The $^{235}_{92}\text{U}$ has undergone α-decay to form $^{231}_{90}$R̶a̶ Th

(iii) The cobalt-60 decays to a nucleus with the same mass number but with one more proton. A neutron has been transformed into a proton and an electron, and the nucleus has undergone β-decay.

Question 2.5

Only rubber gloves or a thin Perspex screen are needed to shield people from plutonium-239, because it undergoes α-decay.

From Appendix 1, the atomic number for plutonium is 94. In α-decay the number of protons (the atomic number, Z) is decreased by two, and so is 92 for the new isotope. The element with $Z = 92$ is uranium.

In α-decay the mass number, A, is reduced by four, so the mass number for the new element is 235. The new isotope is therefore uranium-235. This decay can be written as:

$$^{239}_{94}\text{Pu} \longrightarrow \; ^{235}_{92}\text{U} \; + \; ^{4}_{2}\text{He}$$

Question 2.6

We first find out how many half-lives have elapsed for each of the isotopes.

In 4.5×10^9 years the uranium-235 will have decayed by $(4.5 \times 10^9 / 7.5 \times 10^8)$ half-lives—that is, by 6 half-lives.

The uranium-238 will have decayed by $(4.5 \times 10^9 / 4.5 \times 10^9)$ half-lives, that is by one half-life.

Using Equation 2.7, the fraction of any sample remaining is given by $\frac{1}{2}^n$, where n is the number of half-lives which have elapsed.

For uranium-235, six half-lives have elapsed, so that the fraction of the atoms remaining is $\frac{1}{2}^6 = (1/64) = 0.016$.

For uranium-238, on the other hand, the fraction remaining is $\frac{1}{2}^1$, that is one-half.

Question 2.7

Figure 2.3 plots $(A - Z)$ versus Z. For $^{236}_{92}\text{U}$, $(A - Z) = (236 - 92) = 144$, and $Z = 92$. For $^{140}_{54}\text{Xe}$, $(A - Z) = (140 - 54) = 86$ and $Z = 54$.

For $^{93}_{38}\text{Sr}$, $(A - Z)$ is $(93 - 38) = 55$ and $Z = 38$.

The positions of uranium-236, $^{140}_{54}\text{Xe}$ and $^{93}_{38}\text{Sr}$ are shown in colour on Figure 2.14.

(a) The new nuclei are likely to be unstable, because they lie outside the range of combinations of A and Z for the stable nuclei.

(b) They are unstable because they have too many neutrons for stability.

(c) Because they have too many neutrons for stability, the new nuclei are likely to undergo β-decay, in which a neutron decays into a proton and a β-particle (a high-energy electron), which is emitted from the nucleus.

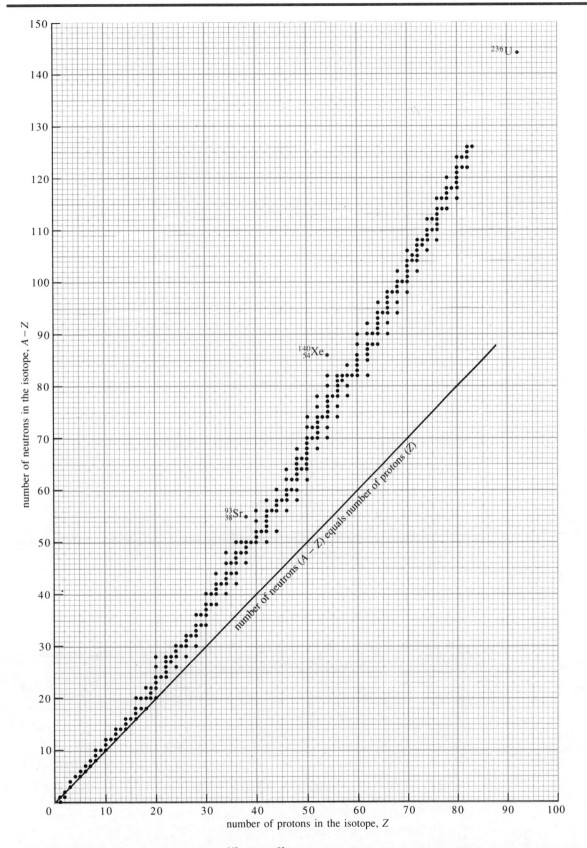

Figure 2.14 The positions of uranium-236, $^{140}_{54}$Xe and $^{93}_{38}$Sr on the graph of the number of protons against the number of neutrons in stable isotopes (Figure 2.3).

(d) The neutron-to-proton ratio in the fission products must, on average, be very similar to that in the parent uranium nucleus. But because the neutron-to-proton ratio for stable isotopes increases with Z, the ratio of about 1.5 which gives a nearly stable nucleus in the uranium region will be too high for stability in the $Z = 30$–60 region, where the fission products occur. Consequently, the fission products will have too many neutrons for stability, and they will undergo β-decay.

Question 2.8

(a) *False*. Fission only occurs in the high-mass elements.

(b) *True*. Neutrons of any energy can cause fission in fissile nuclei.

(c) *True*. Most fission products are radioactive, and undergo β-decay.

Question 2.9

The correct word/phrase to choose for each part is as follows:

(a) less, (b) less, (c) only a limited number, (d) an electron, (e) neutrons of any energy, (f) β-emission, (g) most, (h) less, (i) emission of γ-rays only, (j) very strongly, (k) constant, (l) adding, (m) below, (n) ^{239}Pu.

Question 3.1

From Figure 3.4, the pressure needed to give a boiling temperature of $310\,°C$ is about $100\,atm$, so that the pressure drop needed for the water to boil is $(150 - 100)$ or $50\,atm$. If the pressure drops at $5\,atm$ per minute, the time taken before the coolant boils is therefore $(50/5) = 10$ minutes.

Question 3.2

Rearranging Equation 3.1 gives us:

$$\text{heat produced} = \frac{\text{electrical power out}}{\text{efficiency}/\%} \times 100$$

The PWR has an electrical output of $1\,000\,MW$ and an efficiency of 31%. So,

$$\text{heat produced} = \left(\frac{1\,000}{31} \times 100\right) MW = 3\,226\,MW$$

The core volume is $\pi r^2 h$, where $r = \dfrac{3.4}{2}$ m and $h = 3.6\,m$:

$$\text{core volume} = \pi \times (1.7)^2 \times 3.6 = 32.7\,m^3$$

This gives a power density of $(3\,226\,MW/32.7\,m^3) = 99\,MW\,m^{-3}$ or $99\,kW\,l^{-1}$, which is about 40 times that of an AGR.

Question 3.3

After 1 doubling time the excess fuel from one breeder reactor can be used to fuel a second one; the number of reactors which can be operated therefore doubles (hence the term *doubling time*).

So, after two doubling times, each of the two reactors can be used to fuel another reactor; there would then be 2×2 reactors. After 3 doubling times there would be $(2 \times 2 \times 2)$, and so on. In general, after n doubling times there would be 2^n reactors.

(a) If the doubling time is 10 years, then a 40-year period represents 4 doubling times $(40/10)$. There would therefore be 2^4 reactors at the end of this period, that is, 16 reactors. Each has an output of $1\,000\,MW$ output, so that the total electrical output after forty years would be $16\,000\,MW$.

(b) For a doubling time of 20 years there would be 2 doubling times in 40 years. There would therefore be 2^2 reactors after 40 years. The electricity output would therefore be $2^2 \times 1\,000$ MW, or $4\,000$ MW.

Question 4.1

(a) *True*. The absorbed dose is measured in grays; 1 gray = 1 joule absorbed per kg of tissue.

(b) *True*. The relative biological effectiveness of different radiations compares the absorbed doses needed to produce a particular biological effect.

(c) *False*. An allowance is made by including the quality factor.

(d) *True*. An example of summation is given in Section 4.3.

Question 5.1

Equation 2.7 gives the fraction of atoms remaining after n half-lives as $(1/2)^n$. The radon takes 7.6 days to diffuse to the surface. Expressed in half-lives, this is (7.6 days/3.8 days) = 2 half-lives.

Using Equation 2.7, the fraction that remains is $(1/2)^2 = 0.25$. This corresponds to $0.25 \times 100\%$, that is, 25%.

(*Note*   This means that, as a result of the time it took for the radon to diffuse, the potential dose from the radon would only be 25% of what it would have been had the decay taken place on the surface.)

Question 5.2

The annual collective dose for members of the public from the operation of a PWR at Wylfa was given (p. 94) as 0.043 man Sv yr^{-1}. The estimated collective dose required to produce a fatal cancer was given as 20 man Sv.

In 30 years the collective dose to the public would be

$$0.043\,\text{man Sv yr}^{-1} \times 30\,\text{yr} = 1.29\,\text{man Sv}$$

The estimated number of additional cancers is therefore (1.29 man Sv/20 man Sv) or 0.065 additional cancer deaths as a result of the 30 years' operation of the PWR.

So, much less than one additional death would be expected among the *general UK population* from operation of a PWR at Wylfa for 30 years.

Question 5.3

From Figure 5.14, the initial activity of the fuel is about 10^{20} Bq. After 10, 100 and $1\,000$ years it will have decayed to about 2×10^{17} Bq, 2×10^{16} Bq and 4×10^{13} Bq, respectively. The fractions of the original activity will therefore be $(2 \times 10^{17}/10^{20})$ or 1/500 after 10 years, $(2 \times 10^{16}/10^{20})$ or $1/5\,000$ after 100 years, and $(4 \times 10^{13}/10^{20})$ or four ten-millionths after $1\,000$ years.

Question 5.4

In Figure 5.15, the volume of groundwater is given as $60\,000\,000$ km^3, whereas the volume flowing into the oceans is $20\,000$ km^3, which is $(20\,000/60\,000\,000) \times 100\%$ of the groundwater, or 0.03%.

The volume flowing into the oceans is thus a very small percentage of the total groundwater. The implication of this is that groundwater flows are low.

Question 5.5

In the first clay in Table 5.2, the groundwater flow velocity is $0.13\,\mathrm{m\,yr^{-1}}$. If this mobilizes radioactive waste and carries it to the surface along a path $1\,300\,\mathrm{m}$ long, the time taken for radioactivity to reach the surface will be $(1\,300\,\mathrm{m}/0.13\,\mathrm{m\,yr^{-1}})$ or $10\,000$ years.

Question 6.1

(a) From Figure 6.13, the risk per year that $1\,000$ people will die due to the 100 US nuclear power plants is about 10^{-6}, or 1 in 1 million.

(b) From Figure 6.14, the annual risk that 100 people will die in an air crash is between 1 and 10^{-1}. The annual risk that 100 people will die in a reactor accident under the assumptions made is about 10^{-5}. Thus, the risks from the 100 nuclear power plants would have to be increased by 10^{4} to 10^{5} to equal the risk of 100 people dying in an air crash. So it is between $10\,000$ and $100\,000$ times more likely that 100 people will die in an air crash than in a reactor accident.

(c) From Figure 6.13, the annual risk of $1\,000$ people dying from any natural event is just over 10^{-1}, or 10%. The risk that the same number of people will die due to the 100 nuclear power plants is 10^{-6}. So it is $(10^{-1}/10^{-6})$, that is, $100\,000$ times more likely that this number of people will die annually from natural disasters in the USA than from a reactor accident.

Question 7.1

(a) From Figure 7.1 the minimum weekly demand was about $25\,\mathrm{GW}$ in early August, and the maximum was about $42\,\mathrm{GW}$ in December and January.

(b) From Figure 7.2 the minimum demand on 2 December 1981 was about $22\,\mathrm{GW}$ at about 5 am, and the maximum was about $39\,\mathrm{GW}$ at about 1700 hours.

Question 7.2

The average annual sum of the costs for the nuclear power station is

$£(99 + 38 + 17 + 12)$ million per year $= £166$ million per year

The annual savings are £181 million per year, so the annual net effective cost of the $1\,100\,\mathrm{MW}$ produced by the power station is $(£166\ \mathrm{million} - £181\ \mathrm{million})\,\mathrm{yr^{-1}}$ or $-£15$ million $\mathrm{yr^{-1}}$. Now

$1\,100\,\mathrm{MW} = 1.1 \times 10^{6}\,\mathrm{kW}$

So, expressed as the cost per kW per year:

$$\text{net effective cost} = -\,\frac{£15 \times 10^{6}\,\mathrm{yr^{-1}}}{1.1 \times 10^{6}\,\mathrm{kW}}$$

$$= -£14\,\mathrm{kW^{-1}\,yr^{-1}}$$

The net effective cost of this project is negative, and, based on this method of appraisal, it is therefore worth while to build the new nuclear power station.

Question 7.3

From Figure 7.4 the average UKAEA expenditure from 1975 to 1987 was about £440 million annually (in 1986 money terms), so that 1/4 of this is £110 million. (Your estimate might be different from this, depending on how you found the average. We simply did it by eye, which is quite adequate for this type of discussion.)

Now $1\,\mathrm{TW\,h}$ is $10^{12}\,\mathrm{W\,h}$, so that $42.5\,\mathrm{TW\,h}$ is $42.5 \times 10^{9}\,\mathrm{kW\,h}$. The hidden cost of nuclear power is thus: $(£110 \times 10^{6}/42.5 \times 10^{9})$, or $£0.002\,6\,(\mathrm{kW\,h})^{-1}$, which is $0.26\,\mathrm{p}\,(\mathrm{kW\,h})^{-1}$.

Domestic consumers paid about $6.0\,\mathrm{p\,(kW\,h)^{-1}}$ in 1989 so that, using the assumptions made, the hidden subsidy to them was $(0.26/6.0) \times 100\%$, which is roughly 4%.

Question 8.1

Fission bombs are made from either ^{235}U or ^{239}Pu. In the first of these alternatives, the fissile material is made from natural uranium by means of an enrichment plant; in the second, it is obtained by the reprocessing of reactor fuel of low burn-up. Thus, plutonium is not *enriched*, but is obtained by *reprocessing*.

Question 9.1

A tonne of coal has an energy content of $24 \times 10^9\,\mathrm{J}$, and a tonne of oil has an energy content of $42 \times 10^9\,\mathrm{J}$. Thus, a tonne of oil has as much energy as $\dfrac{42 \times 10^9}{24 \times 10^9}$ t of coal, that is, 1.8 t of coal.

Question 9.2

Figure 9.6 includes an 'all fuels' plot which suggests that in 1960, the fuel input into electricity generation was about 37 million tonnes of oil equivalent, and in 1990, about 70 million tonnes. Thus, in 1960, the estimated efficiency of electricity generation was about $(10/37) \times 100\%$ or 27%; in 1990, about $(25/70) \times 100\%$ or 36%. Thus, following the energy supply crises of the 1970s, there has been a significant increase in the efficiency of electricity generation.

Answers to activities

Activity 1.1

If you are *for* nuclear power you might have included in your list of reasons that:

nuclear power produces less pollution than fossil fuels;

it uses a resource (uranium) that has no alternative uses;

it leaves very little waste to be disposed of;

the amount of material that has to be mined is much less than for coal;

everything has risks, and nuclear power is no worse than the alternatives, and may be a lot better;

any risks that there are, can be reduced by using modern technology;

coal and other fossil fuels are too valuable for other uses to be simply burnt.

If you are *against* nuclear power your list of reasons might include:

nuclear reactors can be used to produce plutonium for use in nuclear weapons;

nuclear reactor accidents can spread radiation over a very wide area;

any increase in radiation levels is bad for you;

the electricity from nuclear power is more expensive than from other sources;

the radioactive waste products cannot be disposed of safely;

we ought to concentrate on renewable sources of energy.

If you are *ambivalent* on the issue, then you will probably have a mixture of these reasons, and may not be able to decide which are the most important. Or you may feel that you are not competent to judge, because the issues are too complex.

Activity 2.1

(a) A ^{1}H nucleus consists of a single proton, and so has a very similar mass to a neutron. By contrast, a ^{238}U nucleus, which contains 238 nucleons, is much more massive. Consequently, neutron–^{1}H collisions correspond to situation (ii) in the model of Figure 2.9, and neutron–^{238}U collisions to situation (i). Thus, after the same number of collisions, the neutron colliding with ^{1}H nuclei will have a much lower speed and energy than the neutron colliding with the ^{238}U nuclei. To give an example, detailed calculations show that, on average, after 18 collisions with ^{1}H nuclei, a 2 MeV neutron has an energy of only about 0.025 eV and is moving at a speed of about 2 200 m s^{-1}. After the 18 collisions with ^{238}U nuclei, the energy of the neutron would be 1.72 MeV; that is, there will have been no appreciable change from the initial neutron energy.

(b) The average loss in energy decreases as the mass of the scattering nucleus increases. The answer can therefore be obtained by arranging the elements in order of increasing mass number. (A good approximate indicator of the mass number of the isotopes can be obtained from the relative atomic masses quoted in Appendix 1.) This order is deuterium (^{2}D), carbon, oxygen, sodium, magnesium, gold and uranium-238. The average loss in neutron energy on collision decreases on moving through this sequence.

(c) The most obvious behaviour which the marble/cannonball model does not reproduce is in cases like those described in Equation 2.2 in which the neutron is absorbed by the nucleus that it encounters. Other examples (including fission) are given later in Section 2.3.2. Cannon-balls do not absorb marbles that are thrown at them; neither do they fracture on impact!

Activity 2.2

(a) As natural uranium contains 99.28% ^{238}U and 0.72% ^{235}U, there are about 7 ^{235}U nuclei present for every 993 ^{238}U nuclei, a ratio of about 1:140. So, if every ^{238}U nucleus could be converted to fissile ^{239}Pu, the amount of fissile material would be expanded 140-fold.

(b) The three stages by which ^{238}U is converted to ^{239}Pu are neutron absorption followed by two successive β-decays. Using the fact that β-decay increases the atomic number by one unit, and that the element with atomic number 91 is protactinium, Pa, the same sequence for $^{232}_{90}$Th gives:

$$^{232}_{90}\text{Th} + {}^{1}_{0}\text{n} \longrightarrow {}^{233}_{90}\text{Th} \xrightarrow{\beta\text{-decay}} {}^{233}_{91}\text{Pa} \xrightarrow{\beta\text{-decay}} {}^{233}_{92}\text{U}$$

In Section 2.3.1, $^{233}_{92}$U is named as an important example of a fissile isotope; in fact it has a half-life of 1.59×10^5 years, and so, like ^{239}Pu, is long lived. If ^{232}Th is four times as plentiful as ^{238}U, this would increase the amount of fissile material by a further factor of 4×140 or 560. Thus, conversion of ^{238}U to ^{239}Pu, and of ^{232}Th to ^{233}U, can in principle increase the amount of fissile material some (140 + 560) or 700-fold.

Activity 3.1

(a) Since the fuel is a compound of natural uranium, the fissile isotope that undergoes fission must be the fissile isotope present in natural uranium. This is ^{235}U. This point is covered in Section 3.1.1.

(b) The fuel is cheap because it is made from *natural* uranium, which contains only 0.72% ^{235}U. In most other reactors the fuel must undergo an expensive process of enrichment to ^{235}U levels of 1.5–4.0% (Section 3.1.1). This is true, for example, of the fuel used in the British AGR (Section 3.2.1), the American PWR (Section 3.2.2) and the Russian RBMK reactor (Section 3.2.3). Over half of the world's reactors are of the PWR type (Section 3.2.2). In the few reactors that do not rely on ^{235}U, the fissile isotope is ^{239}Pu, which must be made by a different but still expensive process of conversion of ^{238}U in a nuclear reactor (Section 3.4).

(c) The fuel is surrounded by D_2O, which is a moderator, as indicated in Figure 3.14 (Section 3.1.3). The CANDU reactor is therefore a thermal reactor.

(d) The coolant passes up the outside of the fuel rods and carries heat away to a steam generator (Section 3.1.2). The substance that does this is D_2O, heavy water. It is not, however, D_2O that is converted to steam for the turbine; the hot D_2O is used to boil H_2O in a steam generator, and it is this which provides the steam. The reactor therefore uses an indirect steam cycle (Section 3.1.2 and Figure 3.3b).

(e) The steam temperature must be very much greater than about 100 °C, the boiling temperature of both H_2O and D_2O at normal pressures, if the efficiency of electrical generation is to be acceptable (Section 3). This will only be so if the liquid D_2O that carries heat from the core to the steam generator has a temperature much greater than 100 °C. As you can see, it is in fact at 293 °C when is enters the steam generator, and at 250 °C when it leaves it. The D_2O can only be in the liquid state at such temperatures if it is under pressure (Section 3.1.2).

(f) Boron and cadmium are good neutron absorbers (Section 3.1.6), and are suitable materials for use in control rods (Figure 3.14), as in this case.

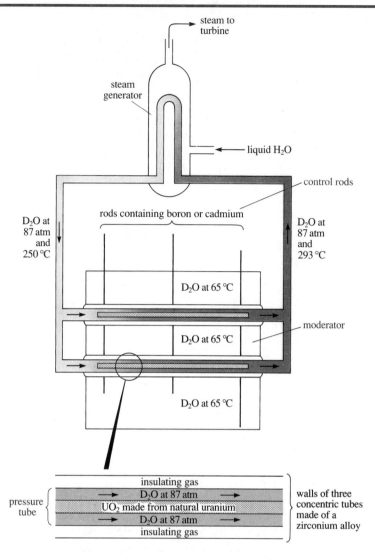

Figure 3.14 Schematic section through the CANDU reactor, showing the locations of the moderator, cladding and materials controlling the chain reaction.

(g) Because so much use is made of D_2O, all of the materials in the core apart from uranium have a low tendency to absorb neutrons. This is why a fuel made from *natural* uranium with its low percentage of ^{235}U can sustain a chain reaction in this system (Section 3.1.1). Reactors containing ordinary light water must use enriched uranium fuel (Section 3.1.1).

(h) A pressure tube is the channel through which the coolant flows under pressure and in which it makes contact with the hot fuel (Section 3.2.3). It is the middle one of the three zirconium alloy tubes in the detail (Figure 3.14). Its purpose is to raise the boiling temperature of the D_2O coolant in order to increase the efficiency of electricity generation.

(i) Substituting into Equation 3.1:

$$28\% = \frac{600 \, \text{MW}}{\text{heat produced}} \times 100\%$$

$$\text{heat produced} = 600 \, \text{MW} \times \frac{100}{28} = 2 \, 143 \, \text{MW}$$

$$\text{core volume} = \pi \times (3.25 \, \text{m})^2 \times 6.0 \, \text{m} = 199 \, \text{m}^3$$

$$\text{power density} = \frac{2 \, 143 \, \text{MW}}{199 \, \text{m}^3} = 10.8 \, \text{MW m}^{-3}$$

This value is well above that of the AGR and RBMK reactors, but well below that of a PWR (Table 3.2).

Activity 3.2

(a) In Table 3.1, the lifetime of uranium resources is estimated to be 62 years at 1987 use rates. This suggests an exhaustion date of 2049. In the bar chart of Extract 3.1, the exhaustion date is 2000.

(b) As the footnote to Table 3.1 makes clear, the estimate assumes that the use of uranium fuel will be confined to the consumption of ^{235}U in thermal reactors. The short uranium lifetime in Extract 3.1 shows that this assumption must have been made in constructing the bar chart. However, uranium could, in principle, make a much greater contribution to energy production by (i) conversion of ^{238}U to fissile ^{239}Pu and consuming this (Section 3.4). This is not mentioned in Extract 3.1. There is a further possibility (ii), which is also not mentioned in the extract, of converting ^{232}Th into fissile ^{233}U, and consuming this (Section 3.3).

(c) In Section 3.3, factor (i) above was said, in practice, to be capable of expanding the quantity of fissile material 60-fold, and factor (ii) to be capable of at least doubling this. Taking a factor of $60 \times 2 = 120$ to be the total expansion, this takes the 20-year lifetime from 1980 in Extract 3.1 up to a 2 400-year lifetime, giving an expiry date of about 4400.

Activity 4.1

(a) From Figure 4.1, the different contributions to the average radiation exposure, in order of *increasing* contribution to the dose are: nuclear power discharges, nuclear weapons fallout, medical uses of radiation and radioisotopes, cosmic rays, natural radioactive substances in food and drink, external radiation from rocks and soils, inhaled radioactive gas from natural sources.

(b) From Figure 4.1 the total average annual radiation dose is 2 150 μSv, of which 1 870 μSv (800 + 400 + 370 + 300 μSv) is from natural sources. This represents (1 870 μSv/2 150 μSv) × 100%, that is 87%.

As you know from buying clothes 'off the peg', there is no 'Mr/Ms Average'. In the same way, the levels of radiation exposure given in Figure 4.1 do not relate to exposure levels that any *particular* individual experiences. Some people may receive radiation exposures much lower than the average, and others may experience higher levels.

For example, you may live in an area like Scotland and Cornwall, where there is a bedrock of granite, which is naturally quite radioactive because of its uranium content. You may live close to a nuclear power station or even live in an area high above sea-level, with increased levels of cosmic radiation. However, the most important variation in our individual exposure levels, as members of the public, is probably due to the extent to which we have been subjected to X-rays for medical or dental reasons.

Activity 4.2

Activity 4.1 showed the average radiation exposure to be 2 150 μSv per year, or about 2 mSv yr^{-1} (0.002 Sv yr^{-1}). Table 4.2 shows that a dose of 1 mSv gives a risk of developing fatal cancer of 1 in 80 000. However, this risk relates to an *instantaneous* dose, and we have already seen that, if a radiation dose is given over a prolonged period, any damage caused has a chance to be repaired. Therefore, our radiation dose of 2 mSv yr^{-1} actually confers a risk of developing cancer of less than 1 in 80 000.

Activity 4.3

(a) The percentage of the exposed population who died from cancer between 1950 and 1982 is:

$$\frac{3\,832}{54\,000} \times 100 = 7.1\%$$

(b) The percentage of the population who would have died from cancer between 1950 and 1982 if the atomic bombs had not been dropped is:

$$\frac{3\,601}{54\,000} \times 100 = 6.7\%$$

(c) The percentage increase in cancer deaths that could be attributed to radiation is:

$$\frac{3\,832 - 3\,601}{3\,601} \times 100 = 6.4\%$$

This represents 231 cancer deaths, which, as a percentage of the whole population is:

$$\frac{231}{54\,000} \times 100 = 0.4\%$$

Activity 4.4

(a) The radiation dose received by the male workers, which was associated with an increased chance of leukaemia in their children, was 100 mSv or more.

(b) At radiation doses above 100 mSv, the risk of their children developing leukaemia was increased by a factor of between 6 and 8.

(c) The other factor investigated as a possible cause of the increased risk of leukaemia was environmental contamination from Sellafield discharges, which could have an effect through eating contaminated food, or playing on a beach.

(d) The evidence that might be in conflict with that from the Sellafield study is that from the atomic bomb survivors, where no excess leukaemias have been observed in the offspring of the men who were irradiated, despite the fact that their external radiation doses were four times higher than those of the Sellafield workers. (The type of radiation received in the two situations may also be relevant, as you see in Activity 4.5.)

(e) The possibility that contamination of their homes might be a factor needs to be investigated further, as does their internal exposure to radiation.

Activity 4.5

(a) Atomic bomb survivors will have been exposed to high *external* radiation doses consisting mainly of γ-rays and neutrons. The incidence of leukaemias in their children is not (on the present evidence) unusually high. However, it may be that the key step in the production of leukaemias in the offspring is the *uptake* of α-emitters such as plutonium by the workers themselves, and that *internal* radiation exposure of this type explains the higher incidence of leukaemias in their children. This is consistent with the fact that Sellafield workers are more exposed to such α-emitters, and more at risk of taking them up than were bomb survivors.

(b) Although α-particles are not very penetrating, and can be stopped by a thin sheet of Perspex, paper or skin, they are densely ionizing, and give one of the *highest* doses—both in terms of the energy transferred per unit mass of tissue (that is, the dose in grays), and taking into account their biological effectiveness (the dose in sieverts). Thus, if α-particles circumvent the skin, and enter the body by inhalation or ingestion, they may be very damaging.

Activity 5.1

(a) The output percentage from this second centrifuge will be the input percentage multiplied by 1.3. This is (0.94% × 1.3) = 1.2%.

(b) The output percentage from the third centrifuge will be the input percentage multiplied by 1.3. This is (1.2% × 1.3) = 1.6%. This value is to be compared with the input percentage into the first centrifuge of 0.72%. It has been obtained by multiplying 0.72% by 1.3, three times in succession: $0.72\% \times 1.3 \times 1.3 \times 1.3 = 1.6\%$. In other words, $1.6\% = 0.72\% \times 1.3^3$.

(c) In (b), the output percentage from the third in a line of three centrifuges was given by the input percentage into the first centrifuge, 0.72%, multiplied by the enrichment factor, 1.3, raised to the power three, three being the number of centrifuges (stages). In general, then, the output percentages, P_f, from the nth stage in a line will be the input percentage into the first stage, P_i, multiplied by the degree of enrichment, α, raised to the power n:

$$P_f = P_i \alpha^n$$

(d) (i) Here $\alpha = 1.3$, $P_i = 0.720\%$ and $n = 6$:

$$P_f = 0.720\% \times 1.3^6 = 3.5\%$$

(ii) Here $\alpha = 1.3$, $P_i = 0.720\%$ and $n = 18$:

$$P_f = 0.720\% \times 1.3^{18} = 81.0\%$$

(iii) Here $\alpha = 1.004\,3$, $P_i = 0.720\%$ and $n = 18$:

$$P_f = 0.720\% \times 1.004\,3^{18} = 0.78\%$$

(iv) Here $\alpha = 1.004\,3$, $P_i = 0.720\%$ and $n = 1\,100$:

$$P_f = 0.720\% \times 1.004\,3^{1\,100} = 80.7\%$$

Within a single enrichment cascade, the gas centrifuge method is much more effective than gaseous diffusion. Thus, comparing (ii) and (iv), 18 gas centrifuges will enrich the uranium to about 80% ^{235}U; it takes 1 100 gaseous diffusion stages to achieve the same result.

Activity 5.2

(a) The repository discussed in Extract 5.1 is intended only for intermediate and low-level waste. Sections 5.6–5.8 have been concerned with high-level waste disposal.

(b) The extract implies that the hydraulic gradient is much lower at Dounreay than at Sellafield because the terrain is flat, and does not provide a head of water from an elevated outcrop.

The velocities in Table 5.2 can be adjusted to the hydraulic gradient of 0.083 by using Equation 5.9. All you need to do is to multiply them by the number obtained when 0.083 is divided by the hydraulic gradient in the table. The resulting velocities are, reading from the top of Table 5.2, 0.054, 7.8×10^{-6}, 2.7 and $0.027\,\mathrm{m\,yr^{-1}}$. Thus, only the first of the two granites has a flow rate of 'a few metres a year' at this hydraulic gradient.

Activity 6.1

(a) Examples of equipment failure include the failure of the circulating pump for cooling water, A, and the failure of the pressure relief valve D to close.

(b) An example of an inadequate maintenance check was the fact that valve M was left closed by mistake.

(c) An example of operator error was switching off the emergency core cooling pump K.

The investigators of the accident also identified faulty and confusing instrumentation as an important contributor. This made the operators' task more difficult than it need have been.

Activity 6.2

Of course, everyone's answer will be different, but the three following paragraphs contain some of the points you could have made; in a sense each paragraph represents a different point of view.

The operators at both nuclear reactor plants were not properly trained. They also interfered with the safeguards built into the plant. Therefore, if training is properly organized, and the reactor is built in such a way that the operators cannot interfere with the safety systems, nuclear reactors would be a lot safer.

The accident at Three Mile Island was not a serious one; very few people are likely to die as a result. The Chernobyl accident was much more serious, but most of the deaths it will cause will take place over a long time. Accidents like Chernobyl, which occur from time to time, may be no worse in the end than the deaths that occur every year from mining coal. And the risks from burning fossil fuels also have to be taken into account, both the risk to health from diseases like bronchitis and pneumoconiosis, and the possible risk to the whole world from global warming.

Human beings will always make mistakes, and it is never possible to be *sure* that control and safety systems will always work perfectly. Radiation is a very special thing: once it has entered the environment, it will give rise to risks until it decays, and that may be thousands of years. It is not right to expose people in the future to risk from what we do now.

Activity 6.3

(a) The main reasons given by Lord Marshall for a Chernobyl-type accident being unlikely in the UK are:

UK reactors are built to high standards with immensely strong containment;
one in ten staff are concerned with safety;
the operators are highly trained;
several 'layers of safety' are provided;
there are independent inspectors;
the Chernobyl reactor would not meet the UK's safety standards.

(b) No information is given to allow you to compare working practices in the two countries. The *implication* is that UK practices are superior, but this is not substantiated: the underlying assumption seems to be that the fact that the Chernobyl accident happened is evidence of their inferior standards.

(c) The article is crisp and very authoritative. It achieves this by making bold assertions, by not admitting to any problems in the UK reactor scene, and by invoking a feeling of competence. Lord Marshall also invites association with the extensive French nuclear power programme, and makes no mention of the problem of human intervention or error. Finally, he associates nuclear power with prosperity and employment, to make us feel that we would suffer if nuclear power were abandoned, and, by implication, he associates nuclear power with reliability of supply, in comparison with coal-fired electricity generation.

Activity 6.4

(a) The advertisement does not say what the risks from the Chernobyl accident are; it invokes the warning about drinking fresh rainwater as evidence of the danger.

(b) The advertisement lists the fossil fuels produced in the UK as energy sources without any mention of their safety; the *implication* is that, by comparison with nuclear power, their safety need not be questioned.

(c) The advertisement uses Friends of the Earth's views on the dangers from nuclear power, as presented at public inquiries, to demonstrate that they had predicted a Chernobyl-type accident. By association, they link the UK nuclear power scene to the accident. They then make confident assertions about alternative energy sources without mentioning any drawbacks.

In general, although this is an advertisement, it is part of a public relations exercise, just like the article based on Lord Marshall's views was (Extract 6.1). As such, the techniques used were very similar, even though the way in which the views were presented are different.

Activity 6.5

(a) The article has a series of messages, which it puts over simply and clearly. It points out that such an accident could happen anywhere, and that the accident is a blow for technology generally. It draws an analogy between the spread of radioactivity by the accident and by nuclear weapons, in order to reinforce the case against nuclear weapons. It stresses the importance of collaboration for solving the problems that the accident will give rise to, and to try to ensure that a similar one does not happen again. Finally, it urges action of a practical nature, which does not attempt to make propaganda.

(b) The tone of this article is a humane one, and should have served to make people think in a more general way about the implications of the accident. The only slightly jarring note is the attempted defence of the Soviet authorities for their delay in warning the rest of the world about what had happened. The article made a nice antidote to Extracts 6.1 and 6.2 which you have just analysed.

Activity 6.6

The event tree is shown in Figure 6.15. Consider the figure and its caption, comparing them with Figure 6.11.

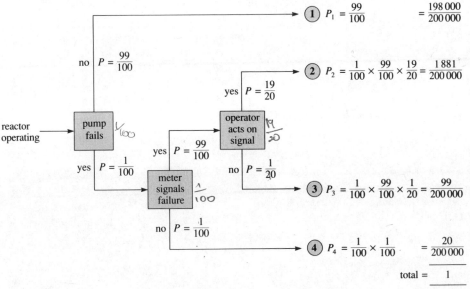

Figure 6.15 The event tree shows a sequence of three events in a nuclear reactor in which a meter either does or does not signal a pump failure, and an operator either does or does not react to the meter signal. As the probabilities of pump failure, meter failure and operator failure are 1/100, 1/100 and 1/20, respectively, those of pump, meter and operator functioning properly are 99/100, 99/100 and 19/20, respectively. Beginning on the left with pump failure, there are four possible outcomes; (i) normal functioning of the pump; (ii) pump failure registered by the meter, which triggers operator action; (iii) pump failure registered by the meter but no response from the operator; (iv) pump failure followed by meter failure. In the figure, these four outcomes are composed of one, three, three and two sequential steps, respectively. On the right, the probability of each outcome, P_1, P_2, P_3 and P_4, is obtained by multiplying together the probabilities of their sequential steps. It is a useful check on the correctness of the calculations to add the four probabilities together, and this is most easily done by converting the four fractions to forms with the same denominator (the lowest common denominator), which in this case is 200 000. The sum of P_1, P_2, P_3 and P_4 is one, since they cover all contingencies. If an accident sequence is defined as one in which the last event is undesirable, the sequences leading to outcomes 3 and 4 are accidents. Their total probability (that is, $P_3 + P_4$) is 119/200 000 or 1/1 680.

Activity 7.1

(a) The original estimate was £4.7 billion, and the latest one £8 billion. The estimated final cost therefore exceeds the original estimate by a factor of (8.0/4.7), that is, 1.7, or 170%.

(b) The reasons given for the latest cost increases were building delays, the cost of fitting out the tunnel and safety modifications.

(c) Income in 1993 is likely to be less than originally estimated because not as many passengers, or as much freight, will be carried. This is because only a limited service could be introduced in the first year, as a result of the safety changes to the trains.

Activity 7.2

(a) If the cost of decommissioning is £1 000 million and it is carried out in year 50, then from Equation 7.2, the present value is given by

$$P = \frac{\text{£1 000 million}}{(1 + d/100)^{50}}$$

At discount rates of 5% and 10%, P would be £87 million and £9 million, respectively. Thus, the four figures in the penultimate row of Table 7.1 become 50, −1 000, −87 and −9, respectively. The net present values at discount rates of 5% and 10% then become £73 million and −£87 million, respectively. Notice that delaying the decommissioning not only reduces the radiation hazard to workers, but it also has a very favourable effect on the net present value. Even though the costs of decommissioning have doubled, the delay to year 50 makes the project even more profitable at a discount rate of 5%, and less unprofitable at a rate of 10%.

(b) The required table is shown as Table 7.2. The discounted annual cash flows in column 4 were calculated by Equation 7.2, with F from columns 2 and 3 and n from column 1. The net present value is just positive. Delaying decommissioning combined with an increase in annual income by 10% (from £150 to £165) have made the project worth while in spite of a doubling of decommissioning costs. The increase in income could come from simply running the plant for a greater percentage of the time, that is, increasing the load factor.

Table 7.2 Discounted cash flow analysis for the nuclear power project of Activity 7.2.

Year of project	Net annual cash flow		Discounted annual cash flow £ million
	cost £ million	income £ million	10% discount rate
0	−1 000		−1 000
1		165	150
2		165	136
3		165	124
4		165	113
5		165	102
6		165	93
7		165	85
8		165	77
9		165	70
10		165	64
50	−1 000		−9
net present value			5

Activity 7.3

(a) There were two problems which arose from the government's privatization proposals. The first was that the private sector needed a higher rate of return to cover what they felt were the risks of the investment. The second was that the priority was given to competition, which meant that the obligation to supply electricity was no longer assured.

(b) The estimated costs increased as a result of:

(i) the requirement to obtain a higher rate of return, 10%, than the government had required (which was 5% and then rose to 8%);

(ii) the need to repay the loan in less than the 35–40 year lifetime of the plant; in particular, the repayment period would be dictated by the length of supply contract which could be obtained, which was likely to be less than 20 years;

(iii) even a 10% rate of return might not be enough to cover the risks that were identified; in particular, the load factor assumed was considered to be too high (75%), and there were uncertainties about decommissioning costs.

(c) The table in the extract shows that the 20-year contract changed the price of electricity from $5.50 \, \mathrm{p} \, (\mathrm{kW \, h})^{-1}$ to $6.25 \, \mathrm{p} \, (\mathrm{kW \, h})^{-1}$, an increase of $0.75 \, \mathrm{p} \, (\mathrm{kW \, h})^{-1}$. This is an increase of $(0.75/5.50) \times 100\%$ or 13.6%.

Activity 8.1

(a) Article I requires the weapons states not to do anything to help the transfer of nuclear weapons or knowledge of them to non-nuclear weapons states. Article II requires the non-weapons states not to receive information or weapons. The non-weapons states also agree to accept a safeguards system run by the IAEA (Article III, Clause 1). Article III, Clause 2 requires countries receiving special fissionable material, or the equipment or material to produce it, to be subject to safeguards.

(b) Article X allows a state to withdraw from the Treaty if 'it decides that extraordinary events … have jeopardized the supreme interests of its country'.

(c) If a country could disguise its nuclear weapons programme as research, then it would be possible for these activities to remain outside the scrutiny of the IAEA.

(d) In our opinion, the essential factors for this Treaty to be successful are (i) effective inspection by the IAEA, (ii) compliance of the countries concerned, and (iii) for all countries to be signatories to the Treaty.

Activity 8.2

(a) Iraq was trying to make a fission weapon containing ^{235}U. The alternative route, which uses ^{239}Pu and relies on the reprocessing of spent reactor fuel, has now been blocked by the bombing of Iraq's nuclear reactors.

(b) Extract 8.3 takes a much more serious view of the Iraqi programme than Extract 8.2, which is quite dismissive. The difference is explained by the discovery that Iraq had access to modern gas centrifuge enrichment technology, rather than to just the old calutron technique. As a result, Iraq may have been within 18 months of having the enriched uranium needed for a weapon, rather than the 30 years estimated in Extract 8.2!

(c) Iraq was clearly in breach of Article II of the Non-proliferation Treaty. It is not clear from Extract 8.2 whether Western governments (and companies) were knowingly in breach of Article I or of Article III, Clause 2. Extract 8.3 implies that the UN and IAEA were reserving judgement on this subject.

As noted at the start of Section 8.4, one of the safeguards referred to in the Non-proliferation Treaty is the IAEA's power of inspection of possible nuclear installations. Clearly the safeguards were not adequate, or were not sufficiently pressed upon the Iraqis, because a £5.8 billion weapons programme went undetected until the outcome of the Gulf War allowed extensive inspections.

However, judgement of the IAEA must be tempered by the recognition that the Non-proliferation Treaty cannot be properly policed without the cooperation of the non-nuclear states. Even under the conditions imposed following the Gulf War, Iraq was still able to conceal many of the documents and materials involved in its weapons programme.

Activity 9.1

(a) Extract 9.1 implies that the production of neutrons within ZETA was taken to indicate the occurrence of a fusion reaction. As the fuel was 100% deuterium, the supposed reaction is one that produces neutrons and occurs between deuterium nuclei. This is given by Equation 9.1:

$$\ce{^2_1H} + \ce{^2_1H} \longrightarrow \ce{^3_2He} + \ce{^1_0n} \tag{9.1}$$

(b) Deuterium (comprising about 0.015% of natural hydrogen) is present in vast quantities in sea-water. (The mass of deuterium in the oceans is about 2.3×10^{13} t. If this underwent the reaction in Equation 9.1 it would produce 1.8×10^{30} J. This would supply the world's energy demand for about 5×10^9 years at 1990 consumption levels!)

Activity 9.2

(a) The JET reaction is that between deuterium and tritium, Equation 9.3:

$$\ce{^2_1H} + \ce{^3_1H} \longrightarrow \ce{^4_2He} + \ce{^1_0n} \tag{9.3}$$

It differs from the reaction that was supposed to have occurred in ZETA in that it involves tritium, and it has an energy output that is over five times greater.

(b) The newspaper report states that fusion occurred over a period of 2 minutes when the actual period was 2 seconds. It also claims that the temperature reached was 300 million degrees centigrade when the press release gives 200 million degrees. Finally, the newspaper report claims that fusion produces no atmospheric pollution. However, you learnt in Section 5.3 that gaseous hydrogen is difficult to contain, so that the radioactive isotope tritium could escape into the atmosphere. In addition, some of the neutrons which are generated will react with the fusion reactor components, producing isotopes that are radioactive.

(c) The inaccuracies in the newspaper article noted in (b) above tend to make the JET experiments seem more sensational, and fusion power more benign, than they actually are. Notice how the erroneously high temperature of 300 million degrees in the newspaper report is said to be 20 times that of the Sun. This implies a Sun temperature of 15 million degrees. In the press release, the correct temperature of 200 million degrees is said to be 10 times that of the Sun's centre, implying a Sun temperature of 20 million degrees. By lowering the Sun temperature, the already erroneous temperature of 300 million degrees given by the newspaper is further sensationalized because it becomes a greater and therefore more impressive multiple of the temperature of the Sun.

Acknowledgements

The Course Team would like to acknowledge the help and advice of the external assessor for this book, Professor S. E. Hunt, and the consultant for Chapter 4, Dr Lillian Somervaille.

Grateful acknowledgement is also made to the following sources for permission to reproduce material in this book:

Text

Extract 3.1 Fraser, A., and Gilchrist, I. (1985), *Starting Science, Book One*, Oxford University Press; *Extract 4.1* Beral, V. (1990), 'Leukaemia and nuclear installations', *British Medical Journal*, **300**, pp. 411–12, British Medical Association; *Extract 4.2* Hawkes, N. (1992), 'Leukaemia clusters linked to low radiation', *The Times*, 20 February 1992, © Times Newspapers Ltd 1992; *Extract 5.1* Wilkie, T. (1992), 'Nuclear waste options narrow', *The Independent on Sunday*, 19 April 1992; 'Nuclear fashion', *The Guardian*, 7 May 1991; *Extract 6.1* Lord Marshall (1986), 'It couldn't happen here', *The People*, 11 May 1986, Syndication International Ltd; *Extract 6.2* 'It need not happen again', © Friends of the Earth (1986); *Extract 6.3* Wilberforce, C. (1986), 'Time for the gloating to stop', *Sunday Mirror*, 11 May 1986, Syndication International Ltd; *Extract 7.1* John, D. (1991), 'Eurotunnel cost rises to £8 billion', *The Guardian*, 8 October 1991; *Extract 7.2* Lord Marshall (1989), 'The facts about nuclear power', *National Power News*, December 1989, National Power; *Extract 8.1* IAEA (1985), 'Treaty on the non-proliferation of nuclear weapons', *International Safeguards and the Non-Proliferation of Nuclear Weapons*, April 1985, International Atomic Energy Agency; *Extract 8.2* Wilkie, T. (1991), 'Scientists explode the myth of Saddam's nuclear bomb', *The Independent on Sunday*, 14 July, 1991; *Extract 8.3* Highfield, R. (1991), 'Iraq planned N-warhead for missiles', *Daily Telegraph*, 5 October 1991, © The Daily Telegraph plc, 1991; Extract 9.1 'Progress towards H-Power', *Manchester Guardian*, 25 January 1958, Guardian Newspapers; *Extract 9.3* Woodman, R., and Wilkie, T. (1991), 'Nuclear fusion a giant step closer', *The Independent on Sunday*, 10 November 1991.

Figures

Figures 5.5, 5.13, 6.3 British Nuclear Fuels plc; *Figure 5.7* British Medical Journal, 7 April 1951, p. 730; *Figure 5.11* Cooper, J. R. (1990), 'The United Kingdom fuel cycle: the radiological impact', *Proc. Conference Power Generation and the Environment,* Institution of Mechanical Engineers, 1990, © National Radiological Protection Board; *Figure 5.21* Blowers, A., Lowry, D., and Solomon, B. (1991), *The International Politics of Nuclear Waste,* Macmillan Ltd (also by permission of St Martin's Press Incorporated); *Figures 6.5, 6.7* Associated Press; *Figure 6.9* Gittus, J. H. *et al.* (1987), *The Chernobyl Accident and its Consequences*, United Kingdom Atomic Energy Authority; *Figure 6.10 The Sun*; *Figures 7.1, 7.2* From the *CEGB Statement of Case to Sizewell B Power Station Public Inquiry*, Appendix H, April 1982, courtesy of Nuclear Electric plc; *Figure 9.1 Daily Telegraph*, 11 November 1991, © The Daily Telegraph plc, 1991; *Figures 9.2, 9.3, 9.4, 9.6, 9.7* Department of Energy (1991), *Digest of UK Energy Statistics*, reproduced with the permission of the Controller of Her Majesty's Stationery Office.

Cartoons

p. 120 Hector Breeze; *p. 142* Solo Syndication; *p. 188* Merrily Harpur.

Index

Note   Entries in **bold** are key terms. Page numbers in *italics* refer to figures and tables.